Ben Stacy Jerrik (Ed.)

Dänikon

AF693911

Ben Stacy Jerrik (Ed.)

Dänikon

Cantons of Switzerland, Canton of Zurich, Dielsdorf district, Communes of Switzerland

Part Press

Imprint

Permission is granted to copy, distribute and/or modify this document under the terms of the GNU Free Documentation License, Version 1.2 or any later version published by the Free Software Foundation; with no Invariant Sections, with the Front-Cover Texts, and with the Back- Cover Texts. A copy of the license is included in the section entitled "GNU Free Documentation License".

All parts of this book are extracted from Wikipedia, the free encyclopedia (www.wikipedia.org).

You can get detailed informations about the authors of this collection of articles at the end of this book. The editors (Ed.) of this book are no authors. They have not modified or extended the original texts.

Pictures published in this book can be under different licences than the GNU Free Documentation License. You can get detailed informations about the authors and licences of pictures at the end of this book.

The content of this book was generated collaboratively by volunteers. Please be advised that nothing found here has necessarily been reviewed by people with the expertise required to provide you with complete, accurate or reliable information. Some information in this book maybe misleading or wrong. The Publisher does not guarantee the validity of the information found here. If you need specific advice (f.e. in fields of medical, legal, financial, or risk management questions) please contact a professional who is licensed or knowledgeable in that area.

Any brand names and product names mentioned in this book are subject to trademark, brand or patent protection and are trademarks or registered trademarks of their respective holders. The use of brand names, product names, common names, trade names, product descriptions etc. even without a particular marking in this works is in no way to be construed to mean that such names may be regarded as unrestricted in respect of trademark and brand protection legislation and could thus be used by anyone.

Cover image: www.ingimage.com
Concerning the licence of the cover image please contact ingimage.

Publisher:
Part Press is a trademark of
International Book Market Service Ltd., 17 Rue Meldrum, Beau Bassin, 1713-01 Mauritius
Email: info@bookmarketservice.com
Website: www.bookmarketservice.com

Published in 2013

Printed in: U.S.A., U.K., Germany. This book was not produced in Mauritius.

ISBN: 978-620-1-55837-3

Contents

Articles

References

Evenkiysky_District

Evenkiysky District (Russian: Эвенки́йский райо́н), or **Evenkia** (Russian: Эвенкия),[1] is an administrative[2] and municipal[3] district (raion), one of the forty-three in Krasnoyarsk Krai, Russia. Before January 1, 2007, it was the only district of Evenk Autonomous Okrug—a federal subject (an autonomous okrug) of Russia. It is located in the central and eastern parts of the krai and borders with Taymyrsky Dolgano-Nenetsky District in the north, the Sakha Republic and Irkutsk Oblast in the east, Kezhemsky, Boguchansky, Motyginsky, and Severo-Yeniseysky Districts in the south, and with Turukhansky District in the west. The area of the district is 763200 square kilometers (sq mi). Its administrative center is the rural locality (a settlement) of Tura.[2] Population: 16,253 (2010 Census);[] 17,697 (2002 Census)[] ; 24,409 (1989 Census).[] The population of Tura accounts for 34.1% of the district's total population.[]

Geography

The southeastern coast of Lake Vivi is a geographical center of Russia.

History

The district was founded on December 4, 2006.[4]

Government

As of 2013, the Head of the district and the Chairman of the District Council is Pyotr I. Suvorov.[5]

Demographics

Vital statistics

Source: Russian Federal State Statistics Service [6]

	Average population (x 1000)	Live births	Deaths	Natural change	Crude birth rate (per 1000)	Crude death rate (per 1000)	Natural change (per 1000)
2007	17	304	233	71	18.1	13.9	4.2
2008	17	290	240	50	17.4	14.4	3.0
2009	16	305	243	62	18.5	14.7	3.8
2010	16	296	213	83	18.1	13.0	5.1

Ethnic groups

Of the 17,697 residents (as of the 2002 Census), 2 (0.01%) chose not to specify their ethnic background. Of the rest, residents identified themselves as belonging to 67 ethnic groups, including ethnic Russians (62%), Evenks (21.5%), Yakuts (5.6%), Ukrainians (3.1%), Kets (1.2%), 162 Tatars (0.9%), 152 Khakas (0.9%) and 127 Volga Germans (0.7%).

Ethnic group	1939 Census		1959 Census		1970 Census		1979 Census		1989 Census		2002 Census	
	Number	%	Number	%	Number	%	Number	%	Number	%	Number	%
Evenks	3,721	39.3%	3,474	33.7%	3,207	25.3%	3,239	20.3%	3,480	14.0%	3,802	21.5%
Yakuts	713	7.5%	51	0.5%	781	6.2%	822	5.1%	937	3.8%	991	5.6%
Kets			14	0.1%	142	1.1%	154	1.0%	150	0.6%	211	1.2%
Russians	4,675	49.4%	5,975	57.9%	7,732	61.1%	10,400	65.1%	16,718	67.5%	10,958	61.9%
Ukrainians	117	1.2%	196	1.9%	254	2.0%	472	3.0%	1,303	5.3%	550	3.1%
Others	234	2.5%	610	5.9%	542	4.3%	881	5.5%	2,181	8.8%	1,185	6.7%

References

Notes

[1] According to Article 2.4 of the Charter of Evenkiysky Municipal District, the names "Evenkiysky Municipal District" and "Evenkia" have equal status.

[2] Law #10-4765

[3] Law #13-6271

[4] Law #652

[5] Official website of Krasnoyarsk Krai. Information about Evenkiysky District (http://www.krskstate.ru/msu/terdel/0/doc/59) (Russian)

[6] http://www.gks.ru/dbscripts/Cbsd/DBInet.cgi#1

Sources

- Законодательное собрание Красноярского края. Закон №10-4765 от 10 июня 2010 г. «О перечне административно-территориальных единиц и территориальных единиц Красноярского края», в ред. Закона №13-6663 от 1 декабря 2011 г. «Об изменении статуса рабочих посёлков и внесении изменений в Закон края "О перечне административно-территориальных единиц и территориальных единиц Красноярского края"». Вступил в силу 1 июля 2010 г.. Опубликован: "Ведомости высших органов государственной власти Красноярского края", №33(404), 5 июля 2010 г. (Legislative Assembly of Krasnoyarsk Krai. Law #10-4765 of June 10, 2010 *On the Registry of the Administrative-Territorial Units and the Territorial Units of Krasnoyarsk Krai*, as amended by the Law #13-6663 of December 1, 2011 *On Changing the Status of the Work Settlements and on Amending the Krai Law "On the Registry of the Administrative-Territorial Units and the Territorial Units of Krasnoyarsk Krai"*. Effective as of July 1, 2010.).
- Законодательное собрание Красноярского края. Закон №13-6271 от 6 октября 2011 г. «Об установлении границ муниципального образования Эвенкийский муниципальный район и находящихся в его границах иных муниципальных образований». Вступил в силу через десять дней после официального опубликования. Опубликован: "Ведомости высших органов государственной власти Красноярского края", №56(497), 1 ноября 2011 г. (Legislative Assembly of Krasnoyarsk Krai. Law #13-6271 of October 6, 2011 *On Establishing the Borders of the Municipal Formation of Evenkiysky District and of Other Municipal Formations Within Its Borders*. Effective as of the day which is ten days after the official publication.).
- Районный совет депутатов Эвенкийского муниципального района. Решение №27-3 от 22 декабря 2005 г. «Устав Эвенкийского муниципального района», в ред. Решения №3-1118-12 от 15 декабря 2012 г. «О внесении изменений в Устав Эвенкийского муниципального района». Вступил в силу после официального опубликования. Опубликован: "Эвенкийская жизнь", №4/2, 2 февраля 2006 г. (District Council of Deputies of Evenkiysky Municipal District. Decision #27-3 of December 22, 2005 *Charter of Evenkiysky Municipal District*, as amended by the Decision #3-1118-12 of December 15, 2012 *On Amending the Charter of Evenkiysky Municipal District*. Effective as of after the official publication.).

- Районный Совет депутатов Эвенкийского муниципального района. Решение №246-10 от 12 сентября 2007 г. «Об утверждении Положения о гимне Эвенкийского муниципального района». Вступил в силу со дня официального публикования. Опубликован: "Эвенкийская жизнь", №37, 20 сентября 2007 г. (District Council of Deputies of Evenkiysky Municipal District. Decision #246-10 of September 10, 2007 *On the Adoption of the Provision on the Anthem of Evenkiysky Municipal District*. Effective as of the day of the official publication.).
- Законодательное Собрание (Суглан) Эвенкийского автономного округа. Закон №652 от 22 ноября 2006 г. «Об упразднении административных единиц Байкитский район, Илимпийский район и Тунгусско-Чунский район и образовании в границах Эвенкийского автономного округа административной единицы Эвенкийский район». Вступил в силу по истечении десяти дней со дня официального опубликования. Опубликован: "Эвенкийская жизнь", №46/2, 23 ноября 2006 г. (Legislative Assembly (Suglan) of Evenk Autonomous Okrug. Law #652 of November 22, 2006 *On the Abolishment of the Administrative Units of Baykitsky District, Ilimpiysky District, and Tungussko-Chunsky District and on the Establishment Within the Borders of Evenk Autonomous Okrug the Administrative Unit of Evenkiysky District*. Effective as of after ten days from the official publication.).

Krasnoyarsk_Krai

Krasnoyarsk Krai	
Красноярский край (Russian)	
— Krai —	
 **Flag**	 **Coat of arms**
Anthem: *It's My Krai, Our Motherland*[]	

Coordinates:	
Political status	
Country	Russia
Federal district	Siberian[1]
Economic region	East Siberian[2]
Established	July 12, 1934[3]
Administrative center	Krasnoyarsk
Government (as of August 2010)	
- Governor[4]	Lev Kuznetsov[5]
- Legislature	Legislative Assembly[4]
Statistics	
Area (as of the 2002 Census)[]	
- Total	2339700 km^2 (sq mi)
Area rank	2nd
Population (2010 Census)[]	
- Total	2828187

- **Rank**	13th
- **Density**[6]	1.21 /km^2 (/sq mi)
- **Urban**	76.3%
- **Rural**	23.7%
Time zone(s)	KRAT (UTC+08:00)[7]
ISO 3166-2	RU-KYA
License plates	24
Official languages	Russian[8]
Official website [9]	

Krasnoyarsk Krai (Russian: Красноя́рский край, tr. *Krasnoyarsky kray*, IPA: [krəsnɐˈjærskʲɪj kraj]) is a federal subject of Russia (a krai). It is the second largest federal subject after the Sakha Republic, Russia's largest krai and the third largest subnational governing body by area in the world, occupying an area of 2339700 square kilometers (sq mi), which is 13% of the country's total territory. The administrative center of the krai is the city of Krasnoyarsk. Population: 2,828,187 (2010 Census).[]

Geography

The krai lies in the middle of Siberia, and belongs to the Siberian Federal District, stretching 3,000 km from the Sayan Mountains in the south along the Yenisei River to the Taymyr Peninsula in the north. It borders Tyumen, Tomsk, Irkutsk, and Kemerovo Oblasts, the Republic of Khakassia, the Tuva Republic, and the Sakha Republic, and the Kara Sea and Laptev Sea of the Arctic Ocean in the north.

The krai is located in the basin of the Arctic Ocean; a great number of rivers that flow through the krai all drain into it eventually. The main rivers of the krai are the Yenisei, and its tributaries (from south to north): the Kan, the Angara, the Podkamennaya Tunguska, and the Nizhnyaya Tunguska.

There are also several thousand lakes in the krai. The largest lakes include Beloye, Belyo, Glubokoye, Itat, Khantayskoye, Labas, Lama, Pyasina, Taymyr, and Yessey. The rivers and lakes are rich with fish.

The climate is strongly continental with large temperature variations during the year. For the central and southern regions where most of the krai's population lives, long winters and short, hot summers are characteristic. The territory of Krasnoyarsk krai experiences conditions of three climate belts: Arctic, Subarctic, and moderate. While in the north there are less than 40 days with temperature higher than 10 °C (°F), in the south there are 110–120 such days.

The average temperature in January is −36 °C (− °F) in the north and −18 °C (− °F) in the south. The average temperature in July is +10 °C (°F) in the north and +20 °C (°F) in the south. The annual precipitation is 316 millimeters (in) (up to 1200 millimeters (in) in foothills of the Sayan Mountains). Snow covers the central regions of the krai from early November until late March. The peaks of the Sayans higher than 2,400–2,600 m and those of the Putorana Plateau higher than 1,000–1,300 m are covered with permanent snow. Permafrost is widespread, especially in the north.

The highest point of the krai is Grandiozny Peak in the East Sayan Mountains at an elevation of 2922 meters (ft).

History

According to archeologists the territory of Siberia was settled around 40,000 BCE[10] The grave-mounds and monuments of the Scythian culture in Krasnoyarsk Krai belong to the 7th century BCE and are ones of the oldest in Eurasia. Among other things a prince's grave Kurgan Arshan exposed in 2001 is known.

The Russian settlement of the area (by Cossacks mostly) began in the 17th century. After the construction of the Trans-Siberian Railroad the Russian colonization of the area strongly increased.

During both the Tsarist and the Bolsheviks' regimes the territory of Krasnoyarsk Krai was used as a place of exile of political enemies of current government. The first leaders of the Soviet state Vladimir Lenin and Joseph Stalin were in exile on the territory of the nowadays krai in 1897–1900 and in 1903 respectively. In Stalin's era numerous Gulag camps were in the region.

In 1822, Yenisei Governorate was created with Krasnoyarsk as its administrative center that covered the territory very close to the one of the current krai.

On June 30, 1908, in the basin of the Podkamennaya Tunguska River, there occurred a powerful explosion most likely to have been caused by the air burst of a large meteoroid or comet fragment at an altitude of 5–10 kilometers (3–6 miles) above Earth's surface. The force of the explosion is estimated to be about 10–15 megatons. It flattened more than 2,000 square kilometers (500,000 acres) of pine forest and killed thousands of reindeer.

Krasnoyarsk Krai was created in 1934 after disaggregation of West Siberian and East Siberian Krais and then included Taymyr and Evenk Autonomous Okrugs and Khakas Autonomous Oblast. In 1991, Khakassia separated from the krai and became a republic within the Russian Federation.

On January 1, 2007, following a referendum on the issue held on April 17, 2005, territories of Evenk and Taymyr Autonomous Okrugs were merged into the krai.

Politics

In December 1991, president Boris Yeltsin appointed Arkady Veprev the first governor of Krasnoyarsk Krai. In January 1993 Boris Yeltsin appointed Valery Zubov the second governor of Krasnoyarsk Krai. In the Krasnoyarsk Territory governor elections were called. Zubov was elected in a universal election for a five-year term. The Legislative Assembly of Krasnoyarsk Krai was created as well.

In 1998, Zubov lost in the gubernatorial election to General Aleksandr Lebed, a politician well known in all Russia. But in 2002 Lebed died in a helicopter accident.

In 2002, Alexander Khloponin, the governor of Taymyr Autonomous Okrug and an influential businessman was elected a governor of Krasnoyarsk Krai. In 2007, he was nominated by president Vladimir Putin for re-election, and Khloponin was elected by the legislative assembly for the second term.

In 2010, after Khloponin was promoted to the office of the president's envoy in the North Caucasian Federal District, Lev Kuznetsov, a businessman and politician from Khloponin's circle, became new governor of the region.

The legislative assembly consists of 52 deputies. 22 of them are elected in 22 one-mandate electoral districts by plurality system, 2 in Taymyr, 2 in Evenkia, and 26 are elected by proportional system from the lists offered by political parties.

Krasnoyarsk Krai is represented in the Federation Council of Russia by two senators. In 2007, eight deputies were elected to the State Duma from Krasnoyarsk regional lists of different political parties.

Economy

Over 95% of the cities, a majority of the industrial enterprises, and all of the agriculture are concentrated in the south of the krai.

Natural resources

The krai is among the richest of Russia's regions in natural resources. Eighty percent of the country's nickel, 75% of its cobalt, 70% of its copper, 16% of its coal, and 10% of its gold are extracted here. Krasnoyarsk also produces 20% of the country's timber. More than 95% of Russian resources of platinum and platinoids are concentrated in the krai.

Industry

The krai's major industries are: non-ferrous metallurgy, energy, forestry, chemicals, and oil refining. The major financial industrial groups of Krasnoyarsk Krai are:

- Basic Element (Krasnoyarsk Aluminium Plant, Achinsk Alumina Complex, Krasnoyarsk Pulp and Paper Plant)
- Interros, (Norilsk Nickel)
- Evraz Group (iron mines, timber)
- MDM Group (SUEK, *Siberian Coal and Energy Company*)
- RusHydro (generation of electric power)
- RZhD (railroads)

Power generation

The two most powerful hydroelectric plants in Russia are built on the Yenisei River, one is being built on the Angara, and several are being planned. It makes Krasnoyarsk Krai one of the most important producers of electric energy in Russia, and a desirable location for energy-intensive industries, such as aluminium plants.

Transportation

- Trans-Siberian Railroad (650 km (400 mi) within Krasnoyarsk Krai), Norilsk Railroad (the most northern railroad in the world); North Siberian Railroad is planned.
- Highways: M53 Baikal (Novosibirsk-Krasnoyarsk-Irkutsk), M54 Yenisei (Krasnoyarsk-Kyzyl-Mongolia) R409 Eniseiskiy trakt (Krasnoyarsk-Lesosibirsk-Eniseisk).
- Northern Sea Route and shipping on Yenisei. The main ports are: Dudinka, Igarka, Turukhansk, Yeniseysk, Maklakovo, Strelka, Krasnoyarsk
- 26 airports, including large international one (Yemelyanovo near Krasnoyarsk).
- Two pipes of Irkutsk-Anzhero-Sudzhensk petroleum pipeline.
- Major power transmission lines.

Administrative divisions

Krasnoyarsk Krai consists of 44 districts and 14 towns of district significance. Two of the districts (Evenkiysky and Taymyrsky; the former autonomous okrugs) have special status.

Demographics

Population (including former Taymyr and Evenk Autonomous Okrugs): 2,828,187 (2010 Census);[] 3,023,525 (2002 Census);[] 3,596,260 (1989 Census).[]

Ethnic groups: The population of the krai mostly consists of Russians, and some other peoples of the former Soviet Union. The indigenous Siberian peoples make up no more than 1% of the population.

Krai Administration's building in the center of Krasnoyarsk

The 2010 Census reported the following ethnic composition:[]

- Russians: 91.3%
- Ukrainians: 1.4%
- Tatars: 1.3%
- Germans: 0.8%
- Azeri: 0.6%
- Belarusians: 0.4%
- Chuvash: 0.4%
- others: 2.5%
- 100,621 people were registered from administrative databases, and could not declare an ethnicity. It is estimated that the proportion of ethnicities in this group is the same as that of the declared group.[11]
- Births (Jan-Aug 2009): 25,520 (13.3 per 1000) [12]
- Deaths (Jan-Aug 2009): 25,183 (13.1 per 1000)

As of August 2009, Krasnoyarsk Krai recorded a natural growth of population for the first time in 16 years.[13]

Total fertility rate:

2003 - 1,35 | 2004 - 1,35 | 2005 - 1,31 | 2006 - 1,33 | 2007 - 1,44 | 2008 - 1,55 | 2009 - 1,61 | 2010 - 1,64 | 2011 - 1,64 |[14]

Vital statistics for 2012

- Births: 41 098 (14.5 per 1000)
- Deaths: 36 726 (12.9 per 1000) [15]
- Total fertility rate: 1.76(e)

Note: Data for Total fertility rate (2012) is estimate based on age and sex structure of Krasnoyarsk Krai at the beginning of 2012, number of births in 2012 and fertility structure in previous years.[16] [17]

Demographics for 2007

[18]

District	Pop	Births	Deaths	NG	BR	DR	NGR
Krasnoyarsk Krai	**2,890,350**	**34,206**	**38,470**	**-4,264**	**11.83**	**13.31**	**-0.15%**
Taimirsky Dolgano-Nenetsky	37,768	592	335	257	15.67	8.87	0.68%
Evenkiysky	16,705	304	233	71	18.20	13.95	0.43%
Abansky	24,997	346	419	-73	13.84	16.76	-0.29%
Achinsky	15,918	226	253	-27	14.20	15.89	-0.17%
Balakhtinsky	23,761	281	409	-128	11.83	17.21	-0.54%
Berezovsky	38,527	483	543	-60	12.54	14.09	-0.16%
Birilyussky	11,431	159	228	-69	13.91	19.95	-0.60%
Bogotolsky	11,371	151	233	-82	13.28	20.49	-0.72%
Boguchansky	48,312	585	626	-41	12.11	12.96	-0.08%
Bolshemurtinsky	19,292	207	398	-191	10.73	20.63	-0.99%
Bolsheuluysky	8,540	112	157	-45	13.11	18.38	-0.53%
Dzerzhinsky	15,025	180	298	-118	11.98	19.83	-0.79%
Novoselovsky	15,128	192	226	-34	12.69	14.94	-0.22%
Partizansky	11,003	155	233	-78	14.09	21.18	-0.71%
Pirovsky	8,251	97	125	-28	11.76	15.15	-0.34%
Ribinsky	23,393	309	422	-113	13.21	18.04	-0.48%

Sayansky	13,058	163	235	-72	12.48	18.00	-0.55%
Severo-Yeniseysky	10,907	153	136	17	14.03	12.47	0.16%
Sukhobuzimsky	23,050	287	362	-75	12.45	15.70	-0.33%
Taseyevsky	13,962	161	234	-73	11.53	16.76	-0.52%
Turukhansky	20,736	249	295	-46	12.01	14.23	-0.22%
Tyukhtetsky	9,034	111	197	-86	12.29	21.81	-0.95%
Uzhursky	33,952	541	586	-45	15.93	17.26	-0.13%
Uyarsky	22,255	250	495	-245	11.23	22.24	-1.10%
Idrinsky	14,037	157	252	-95	11.18	17.95	-0.68%
Ilansky	26,436	352	453	-101	13.32	17.14	-0.38%
Irbeysky	18,053	241	300	-59	13.35	16.62	-0.33%
Kazachinsky	11,333	162	191	-29	14.29	16.85	-0.26%
Sharypovsky	17,816	244	295	-51	13.70	16.56	-0.29%
Shushensky	35,372	392	659	-267	11.08	18.63	-0.75%
Krasnoyarsk	905,000	10,585	10,936	-351	11.70	12.08	-0.04%
Achinsk	110,838	1,333	1,702	-369	12.03	15.36	-0.33%
Bogotol	21,997	273	407	-134	12.41	18.50	-0.61%
Borodino	18,759	197	247	-50	10.50	13.17	-0.27%
Divnogorsk	30,968	337	438	-101	10.88	14.14	-0.33%
Yeniseysk	19,086	265	278	-13	13.88	14.57	-0.07%
Zaozyorny	11,359	184	221	-37	16.20	19.46	-0.33%
Kansk	98,965	1,113	1,458	-345	11.25	14.73	-0.35%
Lesosibirsk	64,215	932	1,027	-95	14.51	15.99	-0.15%
Minusinsk	66,770	852	1,141	-289	12.76	17.09	-0.43%
Nazarovo	53,593	568	890	-322	10.60	16.61	-0.60%
Norilsk	206,359	2,402	1,150	1,252	11.64	5.57	0.61%
Sosnovoborsk	30,074	306	275	31	10.17	9.14	0.10%
Sharypovo	38,495	599	583	16	15.56	15.14	0.04%
Yemelyanovsky	45,908	493	633	-140	10.74	13.79	-0.30%
Kansky	26,696	361	425	-64	13.52	15.92	-0.24%
Karatuzsky	16,992	215	307	-92	12.65	18.07	-0.54%
Kezhemsky	24,406	277	300	-23	11.35	12.29	-0.09%
Kozulsky	18,292	225	344	-119	12.30	18.81	-0.65%
Krasnoturansky	16,098	201	247	-46	12.49	15.34	-0.29%
Kuraginsky	51,402	669	851	-182	13.02	16.56	-0.35%
Mansky	17,684	226	365	-139	12.78	20.64	-0.79%
Minusinsky	26,457	339	409	-70	12.81	15.46	-0.26%
Motiginsky	18,152	238	257	-19	13.11	14.16	-0.10%
Nazarovsky	23,609	390	356	34	16.52	15.08	0.14%

Yeniseysky	27,044	353	418	-65	13.05	15.46	-0.24%
Yermakovsky	20,621	310	360	-50	15.03	17.46	-0.24%
Nizhneingashsky	35,886	448	597	-149	12.48	16.64	-0.42%
Other	245,202	1,673	2,020	-347	6.82	8.24	-0.14%

Religion

Religion in Krasnoyarsk Krai (2012)[19] [20]

Russian Orthodox (29.6%)

Unaffiliated Christian (5%)

Other Orthodox (2%)

Muslim (1.5%)

Rodnovery (1%)

Spiritual but not religious (35%)

Atheist and non-religious (15%)

Other and undeclared (10.9%)

As of a 2012 official survey[19] 29.6% of the population of Krasnoyarsk Krai adheres to the Russian Orthodox Church, 5% declares to be generically Christian (excluding Catholic and Protestant), 2% follows other Orthodox Churches, 1.5% is Muslim, 1% follows Rodnovery (Slavic Paganism), 10.9% did not give an answer to the survey. In addition, 35% of the population declares to be "spiritual but not religious" and 15% to be atheist.[19]

Places of interest

In the krai there are forty-five museums, with total number of about one million objects stored there.

The Krasnoyarsk dam is pictured on Russian ten-ruble note and is a popular attraction for tourists.

The National Park Stolby with its famous rocks neighbors the city of Krasnoyarsk is a popular travelling area and one of the main centres of rock-climbing in Russia.

Education

Krasnoyarsk is the site of the Siberian Federal University, one of the four largest educational institutions of Russia.

Nature and ecology

Many important industrial cities of Krasnoyarsk krai, such as Krasnoyarsk, Norilsk, Achinsk, Kansk, Zheleznogorsk, and Minusinsk, suffer from environmental pollution.

See also

- List of Chairmen of the Legislative Assembly of Krasnoyarsk Krai

Notes and references

[1] Президент Российской Федерации. Указ №849 от 13 мая 2000 г. «О полномочном представителе Президента Российской Федерации в федеральном округе». Вступил в силу 13 мая 2000 г. Опубликован: "Собрание законодательства РФ", №20, ст. 2112, 15 мая 2000 г. (President of the Russian Federation. Decree #849 of May 13, 2000 *On the Plenipotentiary Representative of the President of the Russian Federation in a Federal District*. Effective as of May 13, 2000.).

[2] Госстандарт Российской Федерации. №ОК 024-95 27 декабря 1995 г. «Общероссийский классификатор экономических регионов. 2. Экономические районы», в ред. Изменения №5/2001 ОКЭР. (Gosstandart of the Russian Federation. #OK 024-95 December 27, 1995 *Russian Classification of Economic Regions. 2. Economic Regions*, as amended by the Amendment #5/2001 OKER.).

[3] Всероссийский Центральный Исполнительный Комитет. Постановление от 7 декабря 1934 г. «О разукрупнении Западносибирского и Восточносибирского краёв и образовании новых областей в Сибири». (All-Russian Central Executive Committee. Resolution of December 7, 1934 *On the Subdivision of West Siberian and East Siberian Krais and on the Establishment of New Oblasts in Siberia.*).

[4] Charter, Article 46.3

[5] Official website of Krasnoyarsk krai. Lev Vladimirovich Kuznetsov (http://www.krskstate.ru/governor/biograph/0/doc/266) **(Russian)**

[6] The density value was calculated by dividing the population reported by the 2010 Census by the area shown in the "Area" field. Please note that this value may not be accurate as the area specified in the infobox is not necessarily reported for the same year as the population.

[7] Правительство Российской Федерации. Постановление №725 от 31 августа 2011 г. « О составе территорий, образующих каждую часовую зону, и порядке исчисления времени в часовых зонах, а также о признании утратившими силу отдельных Постановлений Правительства Российской Федерации (http://www.consultant.ru/online/base/?req=doc;base=LAW;n=118896)». Вступил в силу по истечении 7 дней после дня официального опубликования. Опубликован: "Российская Газета", №197, 6 сентября 2011 г. (Government of the Russian Federation. Resolution #725 of August 31, 2011 *On the Composition of the Territories Included into Each Time Zone and on the Procedures of Timekeeping in the Time Zones, as Well as on Abrogation of Several Resolutions of the Government of the Russian Federation* (http://translate.google.com/translate?hl=en&sl=ru&tl=en&u=http://www.consultant.ru/online/base/?req=doc;base=LAW;n=118896). Effective as of after 7 days following the day of the official publication.).

[8] Official the whole territory of Russia according to Article 68.1 of the Constitution of Russia.

[9] http://www.krskstate.ru/

[12] http://www.statis.krs.ru/digital/region5/2007/Показатели%20естественного%20движения%20населения.pdf

[14] http://krasstat.gks.ru/wps/wcm/connect/rosstat_ts/krasstat/ru/statistics/population/

[15] http://www.gks.ru/free_doc/2012/demo/edn12-12.htm

[16] http://www.gks.ru/bgd/regl/b12_111/Main.htm

[17] http://www.gks.ru/doc_2010/demo.pdf

[19] Arena - Atlas of Religions and Nationalities in Russia (http://sreda.org/arena). Sreda.org

[20] 2012 Survey Maps (http://c2.kommersant.ru/ISSUES.PHOTO/OGONIOK/2012/034/ogcyhjk2.jpg). "Ogonek", № 34 (5243), 27/08/2012. *Retrieved 24-09-2012*.

Sources

- Законодательное Собрание Красноярского края. №5-1777 5 июня 2008 г. «Устав Красноярского края». (Legislative Assembly of Krasnoyarsk Krai. #5-1777 June 5, 2008 *Charter of Krasnoyarsk Krai.*).

External links

- Putorana Plateau (http://www.nhpfund.org/nominations/putorana.html) at Natural Heritage Protection Fund (http://www.nhpfund.org/)
- Пантелеев В. НА ПЕРЕЛОМЕ: Аркадий Вепрев у руля. Победа Валерия Зубова на губернаторских выборах//Красноярский рабочий.- 2011. - №145.- 19 августа. (http://www.krasrab.com/archive/2011/08/19/12/view_article)

Russia

Russian Federation	

Anthem: "Государственный гимн Российской Федерации" "*Gosudarstvenny gimn Rossiyskoy Federatsii*"(transliteration) "State Anthem of the Russian Federation"	

Capital and largest city	Moscow
Official languages	Russian official throughout the country; 27 other languages co-official in various regions
Demonym	Russian
Government	Federal semi-presidential constitutional republic
- President	Vladimir Putin
- Prime Minister	Dmitry Medvedev
Legislature	Federal Assembly
- Upper house	Federation Council
- Lower house	State Duma
Formation	
- Kievan Rus'	882
- Grand Duchy of Moscow	1283
- Tsardom of Russia	16 January 1547
- Russian Empire	22 October 1721
- Russian SFSR	6 November 1917
- Part of Soviet Union	10 December 1922

-	Russian Federation	25 December 1991
		Area
-	Total	17,098,242 km^2 (1st) 6,592,800 sq mi
-	Water (%)	13[] (including swamps)
		Population
-	2013 estimate	143,400,000[1] (9th)
-	Density	8.4/km^2 (217th) 21.5/sq mi
GDP (PPP)		2012 estimate
-	Total	$ 3.380 Trillion [2]) (PPP; 5th (6th)
-	Per capita	$23,549[3] (43th)
GDP (nominal)		2012 estimate
-	Total	$2.014 trillion[4]) (8th)
-	Per capita	$14,037[5] (50th)
Gini (2008)		42.3[6] medium · 83rd
HDI (2013)		▲ 0.788[] high · 55th
Currency		Russian ruble (RUB)
Time zone		(UTC+3 to +12[a])
Date format		dd.mm.yyyy
Drives on the		right
Calling code		+7
ISO 3166 code		RU
a.	Excluding +5.	

Russia /ˈrʌʃə/ or /ˈrʊʃə/ (Russian: Россия, tr. *Rossiya*, IPA: [rɐˈsʲijə] (listen)), also officially known as the **Russian Federation**[7] (Russian: Российская Федерация, tr. *Rossiyskaya Federatsiya*, IPA: [rɐˈsʲijskəjə fʲɪdʲɪˈrat͡sɨjə] (listen)), is a country in northern Eurasia.[] It is a federal semi-presidential republic, comprising 83 federal subjects. From northwest to southeast, Russia shares land borders with Norway, Finland, Estonia, Latvia, Lithuania and Poland (both with Kaliningrad Oblast), Belarus, Ukraine, Georgia, Azerbaijan, Kazakhstan, China, Mongolia, and North Korea. It shares maritime borders with Japan by the Sea of Okhotsk, and the US state of Alaska across the Bering Strait. At 17075400 square kilometres (sq mi), Russia is the largest country in the world, covering more than one-eighth of the Earth's inhabited land area. Russia is also the world's ninth most populous nation with 143 million people as of 2012.[8] Extending across the whole of northern Asia and much of Europe, Russia spans nine time zones and incorporates a wide range of environments and landforms.

The nation's history began with that of the East Slavs, who emerged as a recognizable group in Europe between the 3rd and 8th centuries AD.[] Founded and ruled by a Varangian warrior elite and their descendants, the medieval state of Rus arose in the 9th century. In 988 it adopted Orthodox Christianity from the Byzantine Empire,[] beginning the synthesis of Byzantine and Slavic cultures that defined Russian culture for the next millennium.[] Rus' ultimately disintegrated into a number of smaller states; most of the Rus' lands were overrun by the Mongol invasion and became tributaries of the nomadic Golden Horde.[9] The Grand Duchy of Moscow gradually reunified the surrounding Russian principalities, achieved independence from the Golden Horde, and came to dominate the cultural and political legacy of Kievan Rus'. By the 18th century, the nation had greatly expanded through conquest, annexation, and exploration to become the Russian Empire, which was the third largest empire in history, stretching from Poland in Europe to Alaska in North America.[10] [11]

Following the Russian Revolution, the Russian Soviet Federative Socialist Republic became the largest and leading constituent of the Soviet Union, the world's first constitutionally socialist state and a recognized superpower,[12] which played a decisive role in the Allied victory in World War II.[13] [14] The Soviet era saw some of the most significant technological achievements of the 20th century, including the world's first spacecraft, and the first astronaut. The Russian Federation became the successor state of the Russian SFSR following the dissolution of the Soviet Union in 1991, and is recognized as the continuing legal personality of the All-Union state.[]

The Russian economy ranking as the eighth largest by nominal GDP and sixth largest by purchasing power parity. Russia's extensive mineral and energy resources, the largest reserves in the world,[15] have made it one of the largest producers of oil and natural gas globally.[16] [17] The country is one of the five recognized nuclear weapons states and possesses the largest stockpile of weapons of mass destruction.[] Russia is a great power and a permanent member of the United Nations Security Council, a member of the G8, G20, the Council of Europe, the Asia-Pacific Economic Cooperation, the Shanghai Cooperation Organisation, the Eurasian Economic Community, the Organisation for Security and Cooperation in Europe (OSCE), the World Trade Organisation (WTO), and is the leading member of the Commonwealth of Independent States.

Etymology

The name *Russia* is derived from Rus, a medieval state populated mostly by the East Slavs. However, this proper name became more prominent in the later history, and the country typically was called by its inhabitants "Русская Земля" (russkaya zemlya) which could be translated as "Russian Land" or "Land of Rus'". In order to distinguish this state from other states derived from it, it is denoted as *Kievan Rus'* by modern historiography. The name *Rus* itself comes from Rus people, a group of Varangians (possibly Swedish Vikings)[18] [19] who founded the state of Rus (Русь).

An old Latin version of the name Rus' was Ruthenia, mostly applied to the western and southern regions of Rus' that were adjacent to Catholic Europe. The current name of the country, Россия (Rossiya), comes from the Byzantine Greek designation of the Kievan Rus', Ρωσσία *Rossía*—spelt Ρωσία (*Rosía* pronounced [roˈsia]) in Modern Greek.[20]

History

Early periods

Further information: Eurasian nomads, Scythia, Bosporan Kingdom, Goths, Khazars, and East Slavs

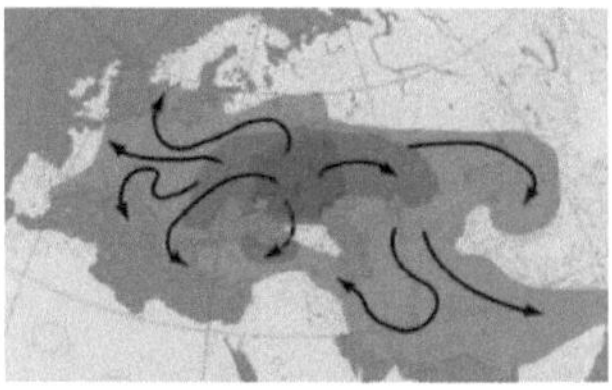

Kurgan hypothesis: South Russia as the urheimat of Indo-European peoples

One of the first modern human bones, dated to 35,000 years old, was found in Russia, in Kostenki on the Don River banks. The only remains of the Denisova hominin that lived about 41,000 years ago were discovered in Denisova Cave (South Siberia).

In prehistoric times the vast steppes of Southern Russia were home to tribes of nomadic pastoralists.[] Remnants of these steppe civilizations were discovered in such places as Ipatovo,[] Sintashta,[21] Arkaim,[22] and Pazyryk,[23] which bear the earliest known traces of mounted warfare, a key feature in the nomadic way of life.

In classical antiquity, the Pontic Steppe was known as Scythia. Since the 8th century BC, Ancient Greek traders brought their civilization to the trade emporiums in Tanais and Phanagoria. The Romans settled on the western part of the Caspian Sea, where their empire stretched towards the east.[24] In 3rd – 4th centuries AD a semi-legendary Gothic kingdom of Oium existed in Southern Russia till it was overrun by Huns. Between the 3rd and 6th centuries AD, the Bosporan Kingdom, a Hellenistic polity which succeeded the Greek colonies,[25] was also overwhelmed by nomadic invasions led by warlike tribes, such as the Huns and Eurasian Avars.[26] A Turkic people, the Khazars, ruled the lower Volga basin steppes between the Caspian and Black Seas until the 10th century.[27]

The ancestors of modern Russians are the Slavic tribes, whose original home is thought by some scholars to have been the wooded areas of the Pinsk Marshes.[28] The East Slavs gradually settled Western Russia in two waves: one moving from Kiev toward present-day Suzdal and Murom and another from Polotsk toward Novgorod and Rostov. From the 7th century onwards, the East Slavs constituted the bulk of the population in Western Russia[29] and slowly but peacefully assimilated the native Finno-Ugric peoples, including the Merya, the Muromians, and the Meshchera.

Kievan Rus

The establishment of the first East Slavic states in the 9th century coincided with the arrival of *Varangians*, the traders, warriors and settlers from the Baltic Sea region. Primarily they were Vikings of Scandinavian origin, who ventured along the waterways extending from the eastern Baltic to the Black and Caspian Seas.[30] According to the Primary Chronicle, a Varangian from Rus' people, named Rurik, was elected ruler of Novgorod in 862. In 882 his successor Oleg ventured south and conquered Kiev,[31] which had been previously paying tribute to the Khazars, founding Kievan Rus'. Oleg, Rurik's son Igor and Igor's son Sviatoslav subsequently subdued all local East Slavic tribes to Kievan rule, destroyed the Khazar khaganate and launched several military expeditions to Byzantium and Persia.

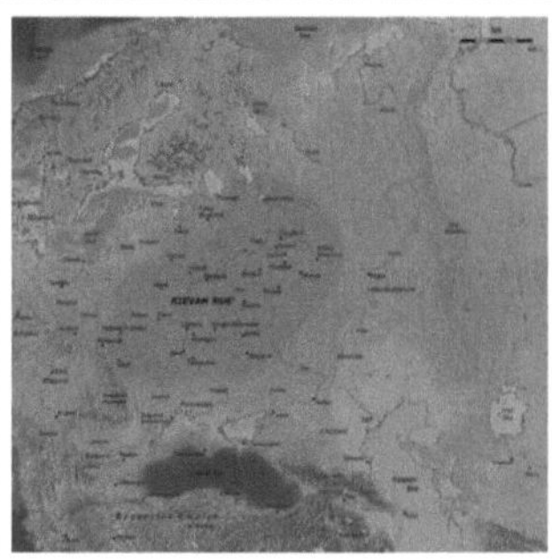

Kievan Rus' in the 11th century

In the 10th to 11th centuries Kievan Rus' became one of the largest and most prosperous states in Europe.[32] The reigns of Vladimir the Great (980–1015) and his son Yaroslav the Wise (1019–1054) constitute the Golden Age of Kiev, which saw the acceptance of Orthodox Christianity from Byzantium and the creation of the first East Slavic written legal code, the *Russkaya Pravda*.

In the 11th and 12th centuries, constant incursions by nomadic Turkic tribes, such as the Kipchaks and the Pechenegs, caused a massive migration of Slavic populations to the safer, heavily forested regions of the north, particularly to the area known as Zalesye.[33]

The Baptism of Kievans, by Klavdy Lebedev

The age of feudalism and decentralization was marked by constant in-fighting between members of the Rurik Dynasty that ruled Kievan Rus' collectively. Kiev's dominance waned, to the benefit of Vladimir-Suzdal in the north-east, Novgorod Republic in the north-west and Galicia-Volhynia in the south-west.

Ultimately Kievan Rus' disintegrated, with the final blow being the Mongol invasion of 1237–40,[34] that resulted in the destruction of Kiev[35] and the death of about half the population of Rus'.[36] The invading Mongol elite, together with their conquered Turkic subjects (Cumans, Kipchaks, Bulgars) became known as Tatars, formed the state of the Golden Horde, which pillaged the Russian principalities; the Mongols ruled the Cuman-Kipchak confederation and Volga Bulgaria (modern-day southern and central expanses of Russia) for over two centuries.[37]

Galicia-Volhynia was eventually assimilated by the Polish-Lithuanian Commonwealth, while the Mongol-dominated Vladimir-Suzdal and Novgorod Republic, two regions on the periphery of Kiev, established the basis for the modern Russian nation.[] The Novgorod together with Pskov retained some degree of autonomy during the time of the Mongol yoke and were largely spared the atrocities that affected the rest of the country. Led by Prince Alexander Nevsky, Novgorodians repelled the invading Swedes in the Battle of the Neva in 1240, as well as the Germanic crusaders in the Battle of the Ice in 1242, breaking their attempts to colonize the Northern Rus'.

Grand Duchy of Moscow

The most powerful successor state to Kievan Rus' was the Grand Duchy of Moscow ("Moscovy" in the Western chronicles), initially a part of Vladimir-Suzdal. While still under the domain of the Mongol-Tatars and with their connivance, Moscow began to assert its influence in the Central Rus' in the early 14th century, gradually becoming the main leading force in the process of the Rus' lands' reunification and expansion of Russia.

Sergius of Radonezh blessing Dmitry Donskoy in Trinity Sergius Lavra, before the Battle of Kulikovo, depicted in a painting by Ernst Lissner

Those were hard times, with frequent Mongol-Tatar raids and agriculture suffering from the beginning of the Little Ice Age. As in the rest of Europe, plagues hit Russia somewhere once every five or six years from 1350 to 1490. However, because of the lower population density and better hygiene (widespread practicing of banya, the wet steam bath),[38] the population loss caused by plagues was not so severe as in the Western Europe, and the pre-Plague populations were reached in Russia as early as 1500.[39]

Led by Prince Dmitry Donskoy of Moscow and helped by the Russian Orthodox Church, the united army of Russian principalities inflicted a milestone defeat on the Mongol-Tatars in the Battle of Kulikovo in 1380. Moscow gradually absorbed the surrounding principalities, including the formerly strong rivals, such as Tver and Novgorod.

Ivan III (*the Great*) finally threw off the control of the Golden Horde, consolidated the whole of Central and Northern Rus' under Moscow's dominion, and was the first to take the title "Grand Duke of all the Russias".[40] After the fall of Constantinople in 1453, Moscow claimed succession to the legacy of the Eastern Roman Empire. Ivan III married Sophia Palaiologina, the niece of the last Byzantine emperor Constantine XI, and made the Byzantine double-headed eagle his own, and eventually Russian, coat-of-arms.

Tsardom of Russia

Tsar Ivan the Terrible by Victor Vasnetsov

In development of the Third Rome ideas, the Grand Duke Ivan IV (the "Terrible"[41]) was officially crowned the first Tsar ("Caesar") of Russia in 1547. The Tsar promulgated a new code of laws (Sudebnik of 1550), established the first Russian feudal representative body (Zemsky Sobor) and introduced local self-management into the rural regions.[42] [43]

During his long reign, Ivan the Terrible nearly doubled the already large Russian territory by annexing the three Tatar khanates (parts of disintegrated Golden Horde): Kazan and Astrakhan along the Volga River, and Sibirean Khanate in South Western Siberia. Thus by the end of the 16th century Russia was transformed into a multiethnic, multidenominational and transcontinental state.

However, the Tsardom was weakened by the long and unsuccessful Livonian War against the coalition of Poland, Lithuania, and Sweden for access to the Baltic coast and sea trade.[44] At the same time the Tatars of the Crimean Khanate, the only remaining successor to the Golden Horde, continued to raid Southern Russia.[45] In effort to restore the Volga khanates, Crimeans and their Ottoman allies invaded central Russia and were even able to burn down parts of Moscow in 1571.[46] But next year the large invading army was thoroughly defeated by Russians in the Battle of Molodi, forever eliminating the threat of the Ottoman-Crimean expansion into Russia. The raids of Crimeans, however, didn't cease until the late 17th century, though the construction of new fortification lines across Southern Russia, such as the Great Abatis Line, constantly narrowed the area accessible to incursions.

Monument to Minin and Pozharsky in Moscow

The death of Ivan's sons marked the end of the ancient Rurik Dynasty in 1598, and in combination with the famine of 1601–03[47] led to the civil war, the rule of pretenders and foreign intervention during the Time of Troubles in the early 17th century.[48] Polish-Lithuanian Commonwealth occupied parts of Russia, including Moscow. In 1612, the Poles were forced to retreat by the Russian volunteer corps, led by two national heroes, merchant Kuzma Minin and Prince Dmitry Pozharsky. The Romanov Dynasty acceded the throne in 1613 by the decision of Zemsky Sobor, and the country started its gradual recovery from the crisis.

Russia continued its territorial growth through the 17th century, which was the age of Cossacks. Cossacks were warriors organized into military communities, resembling pirates and pioneers of the New World. In 1648, the peasants of Ukraine joined the Zaporozhian Cossacks in rebellion against Poland-Lithuania during the Khmelnytsky Uprising, because of the social and religious oppression they suffered under Polish rule. In 1654, the Ukrainian leader, Bohdan Khmelnytsky, offered to place Ukraine under the protection of the Russian Tsar, Aleksey I. Aleksey's acceptance of this offer led to another Russo-Polish War (1654–1667). Finally, Ukraine was split along the Dnieper River, leaving the western part (or Right-bank Ukraine) under Polish rule and eastern part (Left-bank Ukraine and Kiev) under Russian. Later, in 1670–71 the Don Cossacks led by Stenka Razin initiated a major uprising in the Volga Region, but the Tsar's troops were successful in defeating the rebels.

In the east, the rapid Russian exploration and colonisation of the huge territories of Siberia was led mostly by Cossacks hunting for valuable furs and ivory. Russian explorers pushed eastward primarily along the Siberian River Routes, and by the mid-17th century there were Russian settlements in Eastern Siberia, on the Chukchi Peninsula, along the Amur River, and on the Pacific coast. In 1648, the Bering Strait between Asia and North America was passed for the first time by Fedot Popov and Semyon Dezhnyov.

Imperial Russia

Peter the Great, the first Emperor of Russia

Under Peter the Great, Russia was proclaimed an Empire in 1721 and became recognized as a world power. Ruling from 1682 to 1725, Peter defeated Sweden in the Great Northern War, forcing it to cede West Karelia and Ingria (two regions lost by Russia in the Time of Troubles),[49] as well as Estland and Livland, securing Russia's access to the sea and sea trade.[50] On the Baltic Sea Peter founded a new capital called Saint Petersburg, later known as Russia's *Window to Europe*. Peter the Great's reforms brought considerable Western European cultural influences to Russia.

The reign of Peter I's daughter Elizabeth in 1741–62 saw Russia's participation in the Seven Years' War (1756–63). During this conflict Russia annexed East Prussia for a while and even took Berlin. However, upon Elisabeth's death, all these conquests were returned to Kingdom of Prussia by pro-Prussian Peter III of Russia.

Catherine II (*the Great*), who ruled in 1762–96, presided over the Age of Russian Enlightenment. She extended Russian political control over the Polish-Lithuanian Commonwealth and incorporated most of its territories into Russia during the Partitions of Poland, pushing the Russian frontier westward into Central Europe. In the south, after successful Russo-Turkish Wars against the Ottoman Empire, Catherine advanced Russia's boundary to the Black Sea, defeating the Crimean Khanate. As a result of victories over the Ottomans, by the early 19th century Russia also made significant territorial gains in Transcaucasia. This continued with Alexander I's (1801–25) wresting of Finland from the weakened kingdom of Sweden in 1809 and of Bessarabia from the Ottomans in 1812. At the same time Russians colonized Alaska and even founded settlements in California, like Fort Ross.

In 1803–06, the first Russian circumnavigation was made, later followed by other notable Russian sea exploration voyages. In 1820 a Russian expedition discovered the continent of Antarctica.

In alliances with various European countries, Russia fought against Napoleon's France. The French invasion of Russia at the height of Napoleon's power in 1812 failed miserably as the obstinate resistance in combination with the bitterly cold Russian Winter led to a disastrous defeat of invaders, in which more than 95% of the pan-European Grande Armée perished.[51] Led by Mikhail Kutuzov and Barclay de Tolly, the Russian army ousted Napoleon from the country and drove through Europe in the war of the Sixth Coalition, finally entering Paris. Alexander I headed Russia's delegation at the Congress of Vienna that defined the map of post-Napoleonic Europe.

The Russian Empire in 1866 and its spheres of influence

The officers of the Napoleonic Wars brought ideas of liberalism back to Russia with them and attempted to curtail the tsar's powers during the abortive Decembrist revolt of 1825. At the end of the conservative reign of Nicolas I (1825–55), a zenith period of Russia's power and influence in Europe was disrupted by defeat in the Crimean War. Between 1847 and 1851, a massive wave of Asiatic cholera swept over Russia, claiming about one million lives.[52]

Nicholas's successor Alexander II (1855–81) enacted significant changes in the country, including the emancipation reform of 1861. These *Great Reforms* spurred industrialization and modernized the Russian army, which had successfully liberated Bulgaria from Ottoman rule in 1877–78 Russo-Turkish War.

Vladimir Lenin, the leader of the Great October Socialist Revolution

The late 19th century saw the rise of various socialist movements in Russia. Alexander II was killed in 1881 by revolutionary terrorists, and the reign of his son Alexander III (1881–94) was less liberal but more peaceful. The last Russian Emperor, Nicholas II (1894–1917), was unable to prevent the events of the Russian Revolution of 1905, triggered by the unsuccessful Russo-Japanese War and the demonstration incident known as Bloody Sunday. The uprising was put down, but the government was forced to concede major reforms, including granting the freedoms of speech and assembly, the legalization of political parties, and the creation of an elected legislative body, the State Duma of the Russian Empire. Migration to Siberia increased rapidly in the early 20th century, particularly during the Stolypin agrarian reform. Between 1906 and 1914 more than four million settlers arrived in that region.[53]

In 1914, Russia entered World War I in response to Austria-Hungary's declaration of war on Russia's ally Serbia, and fought across multiple fronts while isolated from its Triple Entente allies. In 1916, the Brusilov Offensive of the Russian Army almost completely destroyed the military of Austria-Hungary. However, the already-existing public distrust of the regime was deepened by the rising costs of war, high casualties, and rumors of corruption and treason. All this formed the climate for the Russian Revolution of 1917, carried out in two major acts.

Revolution and Russian Republic

The February Revolution forced Nicholas II to abdicate; he and his family were imprisoned and later executed during the Russian Civil War. The monarchy was replaced by a shaky coalition of political parties that declared itself the Provisional Government. An alternative socialist establishment existed alongside, the Petrograd Soviet, wielding power through the democratically elected councils of workers and peasants, called *Soviets*. The rule of the new authorities only aggravated the crisis in the country, instead of resolving it. Eventually, the October Revolution, led by Bolshevik leader Vladimir Lenin, overthrew the Provisional Government and gave full governing power to the Soviets, leading to the creation of the world's first socialist state.

Soviet Russia and civil war

Following the October Revolution, a civil war broke out between the anti-communist White movement and the new Soviet regime with its Red Army. Bolshevist Russia lost its Ukrainian, Polish, Baltic, and Finnish territories by signing the Treaty of Brest-Litovsk that concluded hostilities with the Central Powers of World War I. The Allied powers launched an unsuccessful military intervention in support of anti-Communist forces. In the meantime both the Bolsheviks and White movement carried out campaigns of deportations and executions against each other, known respectively as the Red Terror and White Terror. By the end of the civil war, the Russian economy and its infrastructure were heavily damaged. Millions became White émigrés,[54] and the Povolzhye famine of 1921 claimed up to 5 million victims.[55]

The symbols of the early Soviet era: Tatlin's Tower project and the giant *Worker and Kolkhoz Woman* sculpture group

Soviet Union

The Russian Soviet Federative Socialist Republic (called *Russian Socialist Federative Soviet Republic* at the time) together with the Ukrainian, Byelorussian, and Transcaucasion Soviet Socialist Republics, formed the Union of Soviet Socialist Republics (USSR), or Soviet Union, on 30 December 1922. Out of the 15 republics that would make up the USSR, the largest in size and over half of the total USSR population was the Russian SFSR, which came to dominate the union for its entire 69-year history.

Following Lenin's death in 1924, a troika was designated to govern the Soviet Union. However, Joseph Stalin, an elected General Secretary of the Communist Party, managed to suppress all opposition groups within the party and consolidate power in his hands. Leon Trotsky, the main proponent of world revolution, was exiled from the Soviet Union in 1929, and Stalin's idea of Socialism in One Country became the primary line. The continued internal struggle in the Bolshevik party culminated in the Great Purge, a period of mass repressions in 1937–38, during which hundreds of thousands of people were executed, including original party members and military leaders accused of coup d'état plots.[56]

The Red Army is greeted in Bucharest, August 1944

Under Stalin's leadership, the government launched a planned economy, industrialisation of the largely rural country, and collectivization of its agriculture. During this period of rapid economic and social change, millions of people were sent to penal labor camps,[57] including many political convicts for their opposition to Stalin's rule; millions were deported and exiled to remote areas of the Soviet Union.[57] The transitional disorganisation of the country's agriculture, combined with the harsh state policies and a drought, led to the Soviet famine of 1932–1933.[58] The Soviet Union, though with a heavy price, was transformed from a largely agrarian economy to a major industrial powerhouse in a short span of time.

The Appeasement policy of Great Britain and France towards Adolf Hitler's annexation of Austria and Czechoslovakia enlarged the might of Nazi Germany and put a threat of war to the Soviet Union. Around the same time the Third Reich allied with the Empire of Japan, a rival of the USSR in the Far East and an open enemy of the USSR in the Soviet–Japanese Border Wars in 1938–39.

In August 1939, after another failure of attempts to establish an anti-Nazi alliance with Britain and France, the Soviet government decided to improve relations with Germany by concluding the Molotov-Ribbentrop Pact, pledging non-aggression between the two countries and dividing their spheres of influence in Eastern Europe. While Hitler conquered Poland, France and other countries actied on single front at the start of World War II, the USSR was able to build up its military and claim some of the former territories of the Russian Empire as a result of the Soviet invasion of Poland, Winter War and the occupation of the Baltic states.

On 22 June 1941, Nazi Germany broke the non-aggression treaty and invaded the Soviet Union with the largest and most powerful invasion force in human history,[59] opening the largest theater of World War II. Although the German army had considerable early success, their attack was halted in the Battle of Moscow. Subsequently the Germans were dealt major defeats first at the Battle of Stalingrad in the winter of 1942–43,[60] and then in the Battle of Kursk in the summer of 1943. Another German failure was the Siege of Leningrad, in which the city was fully blockaded on land between 1941–44 by German and Finnish forces, and suffered starvation and more than a million deaths, but never surrendered.[61] Under Stalin's administration and the leadership of such commanders as Georgy Zhukov and Konstantin Rokossovsky, Soviet forces took Eastern Europe in 1944–45 and captured Berlin in May 1945. In August 1945 the Soviet Army ousted the Japanese from China's Manchukuo and North Korea, contributing to the allied victory over Japan.

The 1941–45 period of World War II is known in Russia as the *Great Patriotic War*. During this conflict, which included many of the most lethal battle operations in human history, Soviet military and civilian deaths were 10.6 million and 15.9 million respectively,[62] accounting for about a third of all World War II casualties. The full demographic loss to the Soviet peoples was even greater.[63] The Soviet economy and infrastructure suffered massive devastation[64] but the Soviet Union emerged as an acknowledged military superpower on the continent.

First human to travel into space, Yuri Gagarin

The Red Army occupied Eastern Europe after the war, including East Germany. Dependent socialist governments were installed in the Eastern Bloc satellite states. Becoming the world's second nuclear weapons power, the USSR established the Warsaw Pact alliance and entered into a struggle for global dominance, known as the Cold War, with the United States and NATO. The Soviet Union supported revolutionary movements across the world, including the newly formed People's Republic of China, the Democratic People's Republic of Korea and, later on, the Republic of Cuba. Significant amounts of the Soviet resources were allocated in aid to the other socialist states.[65]

After Stalin's death and a short period of collective rule, new leader Nikita Khrushchev denounced the cult of personality of Stalin and launched the policy of de-Stalinization. The penal labor system was reformed and many prisoners were released and rehabilitated (many of them posthumously).[66] The general easement of repressive policies became known later as the Khrushchev Thaw. At the same time, tensions with the United States heightened when the two rivals clashed over the deployment of the U.S. Jupiter missiles in Turkey and Soviet missiles in Cuba.

In 1957, the Soviet Union launched the world's first artificial satellite, *Sputnik 1*, thus starting the Space Age. Russian cosmonaut Yuri Gagarin became the first human to orbit the Earth aboard *Vostok 1* manned spacecraft on 12 April 1961.

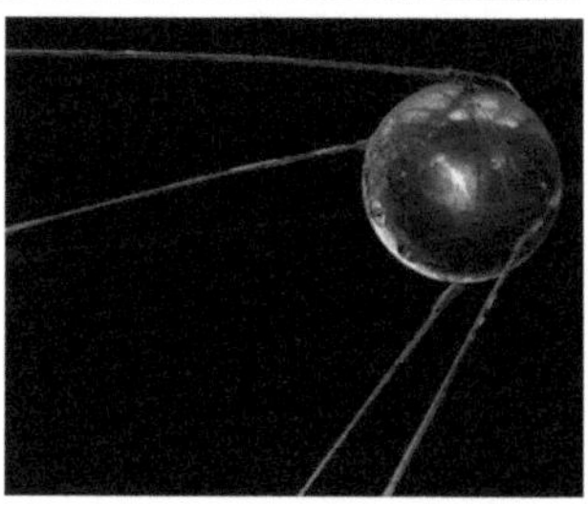

Sputnik 1 was the world's first artificial satellite

Following the ousting of Khrushchev in 1964, another period of collective rule ensued, until Leonid Brezhnev became the leader. The era of 1970s and the early 1980s was designated later as the Era of Stagnation, a period when the economic growth slowed and social policies became static. The 1965 Kosygin reform, aimed for partial decentralization of the Soviet economy and shifted the emphasis from heavy industry and weapons to light industry and consumer goods, but was stifled by the conservative Communist leadership.

In 1979, after a Communist-led revolution in Afghanistan, Soviet forces entered that country at request of the new regime. The occupation drained economic resources and dragged on without achieving meaningful political results. Ultimately the Soviet Army was withdrawn from Afghanistan in 1989 due to international opposition, persistent anti-Soviet guerilla warfare, and a lack of support by Soviet citizens.

From 1985 onwards, the last Soviet leader Mikhail Gorbachev, who sought to enact liberal reforms in the Soviet system, introduced the policies of *glasnost* (openness) and *perestroika* (restructuring) in an attempt to end the period of economic stagnation and to democratise the government. This, however, led to the rise of strong nationalist and separatist movements. Prior to 1991, the Soviet economy was the second largest in the world,[67] but during its last years it was afflicted by shortages of goods in grocery stores, huge budget deficits, and explosive growth in money supply leading to inflation.[68]

By 1991, economic and political turmoil began to boil over, as the Baltic republics chose to secede from the Union. On 17 March, a referendum was held, to which the vast majority of participating citizens voted in favour of preserving the Soviet Union as a renewed federation. In August 1991, a coup d'état attempt by members of Gorbachev's government, directed against Gorbachev and aimed at preserving the Soviet Union, instead led to the end of the Communist Party of the Soviet Union. Despite the will expressed by the people, on 25 December 1991, the USSR was dissolved into 15 post-Soviet states.

Russian Federation

Moscow International Business Center under construction

Boris Yeltsin was elected the President of Russia in June 1991, in the first direct presidential election in Russian history. During and after the Soviet disintegration, wide-ranging reforms including privatization and market and trade liberalization were undertaken,[] including radical changes along the lines of "shock therapy" as recommended by the United States and International Monetary Fund.[69] All this resulted in a major economic crisis, characterized by 50% decline of both GDP and industrial output between 1990–95.[] [70]

The privatization largely shifted control of enterprises from state agencies to individuals with inside connections in the government. Many of the newly rich moved billions in cash and assets outside of the country in an enormous capital flight.[71] The depression of the economy led to the collapse of social services; the birth rate plummeted while the death rate skyrocketed. Millions plunged into poverty, from 1.5% level of poverty in the late Soviet era, to 39–49% by mid-1993.[] The 1990s saw extreme corruption and lawlessness, the rise of criminal gangs and violent crime.[72]

The 1990s were plagued by armed conflicts in the North Caucasus, both local ethnic skirmishes and separatist Islamist insurrections. From the time Chechen separatists declared independence in the early 1990s, an intermittent guerrilla war has been fought between the rebel groups and the Russian military. Terrorist attacks against civilians carried out by separatists, most notably the Moscow theater hostage crisis and Beslan school siege, caused hundreds of deaths and drew worldwide attention.

Russia took up the responsibility for settling the USSR's external debts, even though its population made up just half of the population of the USSR at the time of its dissolution.[73] High budget deficits caused the 1998 Russian financial crisis[74] and resulted in a further GDP decline.[]

On 31 December 1999, President Yeltsin resigned, handing the post to the recently appointed Prime Minister, Vladimir Putin, who then won the 2000 presidential election. Putin suppressed the Chechen insurgency, although sporadic violence still occurs throughout the Northern Caucasus. High oil prices and the initially weak currency followed by increasing domestic demand, consumption and investments has helped the economy to grow for nine straight years, improving the standard of living and increasing Russia's influence on the world stage.[17] While many reforms made during the Putin presidency have been generally criticized by Western nations as un-democratic,[75] Putin's leadership over the return of order, stability, and progress has won him widespread popularity in Russia.[76]

On 2 March 2008, Dmitry Medvedev was elected President of Russia, whilst Putin became Prime Minister. Putin returned to the presidency following the 2012 presidential elections, and Medvedev was appointed Prime Minister.

Politics

Governance

Moscow Kremlin, the working residence of the President of Russia

According to the Constitution of Russia, the country is a federation and semi-presidential republic, wherein the President is the head of state[77] and the Prime Minister is the head of government. The Russian Federation is fundamentally structured as a multi-party representative democracy, with the federal government composed of three branches:

- Legislative: The bicameral Federal Assembly of Russia, made up of the 450-member State Duma and the 166-member Federation Council, adopts federal law, declares war, approves treaties, has the power of the purse and the power of impeachment of the President.
- Executive: The President is the Supreme Commander-in-Chief of the Armed Forces, can veto legislative bills before they become law, and appoints the Government of Russia (Cabinet) and other officers, who administer and enforce federal laws and policies.
- Judiciary: The Constitutional Court, Supreme Court, Supreme Court of Arbitration and lower federal courts, whose judges are appointed by the Federation Council on the recommendation of the President, interpret laws and can overturn laws they deem unconstitutional.

The president is elected by popular vote for a six-year term (eligible for a second term, but not for a third consecutive term).[78] Ministries of the government are composed of the Premier and his deputies, ministers, and selected other individuals; all are appointed by the President on the recommendation of the Prime Minister (whereas the appointment of the latter requires the consent of the State Duma). Leading political parties in Russia include United Russia, the Communist Party, the Liberal Democratic Party of Russia, and A Just Russia.

Foreign relations

Leaders of the BRIC nations in 2008: (l-r) Manmohan Singh of India, Dmitry Medvedev of Russia, Hu Jintao of China and Luiz Inácio Lula da Silva of Brazil

The Russian Federation is recognized in international law as a successor state of the former Soviet Union.[] Russia continues to implement the international commitments of the USSR, and has assumed the USSR's permanent seat in the UN Security Council, membership in other international organisations, the rights and obligations under international treaties, and property and debts. Russia has a multifaceted foreign policy. As of 2009, it maintains diplomatic relations with 191 countries and has 144 embassies. The foreign policy is determined by the President and implemented by the Ministry of Foreign Affairs of Russia.[79]

As the successor to a former superpower, Russia's geopolitical status has been often debated, particularly in relation to unipolar and multipolar views on the global political system. While Russia is commonly accepted to be a great power, in recent years it has been characterized by a number of world leaders,[80] [81] scholars,[82] commentators and politicians[83] as a currently reinstating or potential superpower.[84] [85] [86]

An important aspect of Russia's relations with the West is the criticism of Russia's political system and human rights management by the Western governments, the mass media and the leading democracy and human rights watchdogs.

In particular, such organisations as the Amnesty International and Human Rights Watch consider Russia to have not enough democratic attributes and to allow few political rights and civil liberties to its citizens.[87] [88] Freedom

House, an international organisation funded by the United States, ranks Russia as "not free", citing "carefully engineered elections" and "absence" of debate.[89] Russian authorities dismiss these claims and especially criticise Freedom House. The Russian Ministry of Foreign Affairs has called the 2006 *Freedom in the World* report "prefabricated", stating that the human rights issues have been turned into a political weapon in particular by the United States. The ministry also claims that such organisations as Freedom House and Human Rights Watch use the same scheme of voluntary extrapolation of *"isolated facts that of course can be found in any country"* into *"dominant tendencies"*.[90]

Member states, observers and partners of the Shanghai Cooperation Organisation

As one of five permanent members of the UN Security Council, Russia plays a major role in maintaining international peace and security. The country participates in the Quartet on the Middle East and the Six-party talks with North Korea. Russia is a member of the G8 industrialized nations, the Council of Europe, OSCE and APEC. Russia usually takes a leading role in regional organisations such as the CIS, EurAsEC, CSTO, and the SCO.[91] Former President Vladimir Putin had advocated a strategic partnership with close integration in various dimensions including establishment of EU-Russia Common Spaces.[92] Since the dissolution of the Soviet Union, Russia has developed a friendlier relationship with the United States and NATO. The NATO-Russia Council was established in 2002 to allow the United States, Russia and the 25 allies in NATO to work together as equal partners to pursue opportunities for joint collaboration.[93]

Russia maintains strong and positive relations with other BRIC countries. In recent years, the country has sought to strengthen ties especially with the People's Republic of China by signing the Treaty of Friendship as well as building the Trans-Siberian oil pipeline geared toward growing Chinese energy needs.[94]

Military

The Russian military is divided into the Ground Forces, Navy, and Air Force. There are also three independent arms of service: Strategic Missile Troops, Russian Space Forces, and the Airborne Troops. In 2006, the military had 1.037 million personnel on active duty.[] It is mandatory for all male citizens aged 18–27 to be drafted for a year of service in Armed Forces.[17]

2010 Moscow Victory Day Parade

Russia has the largest stockpile of nuclear weapons in the world. It has the second largest fleet of ballistic missile submarines and is the only country apart from the U.S. with a modern strategic bomber force.[] [95] Russia's tank force is the largest in the world, its surface navy and air force are among the largest ones.

The country has a large and fully indigenous arms industry, producing most of its own military equipment with only few types of weapons imported. Russia is the world's top supplier of arms, a spot it has held since 2001, accounting for around 30% of worldwide weapons sales[96] and exporting weapons to about 80 countries.[97]

Official government military spending for 2012 was $71.9 billion, the third largest in the world, though various sources have estimated Russia's military expenditures to be considerably higher.[] [98] Currently, a major equipment upgrade worth about $200 billion is on its way between 2006 and 2015.[99]

According to 2012 Global Peace Index, Russia is the sixth least peaceful country in the world, principally because of its defense industry. Russia has historically ranked low on the index since its inception in 2007.[100]

Political divisions

Federal subjects

The Russian Federation comprises 83 federal subjects.[101] These subjects have equal representation—two delegates each—in the Federation Council.[102] However, they differ in the degree of autonomy they enjoy.

- 46 oblasts (provinces): most common type of federal subjects, with locally elected[103] governor and legislature.
- 21 republics: nominally autonomous; each has its own constitution, direct-elected[103] head of republic[104] or a similar post, and parliament. Republics are allowed to establish their own official language alongside Russian but are represented by the federal government in international affairs. Republics are meant to be home to specific ethnic minorities.
- 9 krais (territories): essentially the same as oblasts. The "territory" designation is historic, originally given to frontier regions and later also to the administrative divisions that comprised autonomous okrugs or autonomous oblasts.
- 4 autonomous okrugs (autonomous districts): originally autonomous entities within oblasts and krais created for ethnic minorities, their status was elevated to that of federal subjects in the 1990s. With the exception of Chukotka Autonomous Okrug, all autonomous okrugs are still administratively subordinated to a krai or an oblast of which they are a part.
- 1 autonomous oblast (the Jewish Autonomous Oblast): historically, autonomous oblasts were administrative units subordinated to krais. In 1990, all of them except for the Jewish AO were elevated in status to that of a republic.
- 2 federal cities (Moscow and St. Petersburg): major cities that function as separate regions.

Federal districts

Federal subjects are grouped into eight federal districts, each administered by an envoy appointed by the President of Russia.[105] Unlike the federal subjects, the federal districts are not a subnational level of government, but are a level of administration of the federal government. Federal districts' envoys serve as liaisons between the federal subjects and the federal government and are primarily responsible for overseeing the compliance of the federal subjects with the federal laws.

Geography

Russia is the largest country in the world; its total area is 17075400 square kilometres (sq mi). There are 23 UNESCO World Heritage Sites in Russia, 40 UNESCO biosphere reserves,[106] 41 national parks and 101 nature reserves. It lies between latitudes 41° and 82° N, and longitudes 19° E and 169° W.

Russia has a wide natural resource base, including major deposits of timber, petroleum, natural gas, coal, ores and other mineral resources.

The topography of Russia

Topography

The two widest separated points in Russia are about 8000 km (mi) apart along a geodesic line. These points are: the boundary with Poland on a 60 km (mi) long Vistula Spit separating the Gdańsk Bay from the Vistula Lagoon; and the farthest southeast of the Kuril Islands. The points which are furthest separated in longitude are 6600 km (mi) apart along a geodesic line. These points are: in the west, the same spit; in the east, the Big Diomede Island. The Russian Federation spans 9 time zones.

Mount Elbrus, the highest point of the Caucasus, Russia and Europe

Most of Russia consists of vast stretches of plains that are predominantly steppe to the south and heavily forested to the north, with tundra along the northern coast. Russia possesses 10% of the world's arable land.[107] Mountain ranges are found along the southern borders, such as the Caucasus (containing Mount Elbrus, which at 5642 m (ft) is the highest point in both Russia and Europe) and the Altai (containing Mount Belukha, which at the 4506 m (ft) is the highest point of Siberia outside of the Russian Far East); and in the eastern parts, such as the Verkhoyansk Range or the volcanoes of Kamchatka Peninsula (containing Klyuchevskaya Sopka, which at the 4750 m (ft) is the highest active volcano in Eurasia as well as the highest point of Asian Russia). The Ural Mountains, rich in mineral resources, form a north-south range that divides Europe and Asia.

Russia has an extensive coastline of over 37000 km (mi) along the Arctic and Pacific Oceans, as well as along the Baltic Sea, Sea of Azov, Black Sea and Caspian Sea.[17] The Barents Sea, White Sea, Kara Sea, Laptev Sea, East Siberian Sea, Chukchi Sea, Bering Sea, Sea of Okhotsk, and the Sea of Japan are linked to Russia via the Arctic and Pacific. Russia's major islands and archipelagos include Novaya Zemlya, the Franz Josef Land, the Severnaya Zemlya, the New Siberian Islands, Wrangel Island, the Kuril Islands, and Sakhalin. The Diomede Islands (one controlled by Russia, the other by the U.S.) are just 3 km (mi) apart, and Kunashir Island is about 20 km (mi) from Hokkaido, Japan.

Russia has thousands of rivers and inland bodies of water, providing it with one of the world's largest surface water resources. Its lakes contain approximately one-quarter of the world's liquid fresh water.[] The largest and most prominent of Russia's bodies of fresh water is Lake Baikal, the world's deepest, purest, oldest and most capacious fresh water lake.[] Baikal alone contains over one-fifth of the world's fresh surface water.[] Other major lakes include Ladoga and Onega, two of the largest lakes in Europe. Russia is second only to Brazil in volume of the total renewable water resources. Of the country's 100,000 rivers,[108] the Volga is the most famous, not only because it is the longest river in Europe, but also because of its major role in Russian history.[17] The Siberian rivers Ob, Yenisey, Lena and Amur are among the longest rivers in the world.

Central Russian Upland near Zaraysk, Moscow Oblast

Climate

The enormous size of Russia and the remoteness of many areas from the sea result in the dominance of the humid continental climate, which is prevalent in all parts of the country except for the tundra and the extreme southeast. Mountains in the south obstruct the flow of warm air masses from the Indian Ocean, while the plain of the west and north makes the country open to Arctic and Atlantic influences.[]

Taiga forest in winter, Arkhangelsk Oblast

Most of Northern European Russia and Siberia has a subarctic climate, with extremely severe winters in the inner regions of Northeast Siberia (mostly the Sakha Republic, where the Northern Pole of Cold is located with the record low temperature of), and more moderate elsewhere. The strip of land along the shore of the Arctic Ocean, as well as the Russian Arctic islands, have a polar climate.

The coastal part of Krasnodar Krai on the Black Sea, most notably in Sochi, possesses a humid subtropical climate with mild and wet winters. Winter is dry compared to summer in many regions of East Siberia and the Far East, while other parts of the country experience more even precipitation across seasons. Winter precipitation in most parts of the country usually falls as snow. The region along the Lower Volga and Caspian Sea coast, as well as some areas of southernmost Siberia, possesses a semi-arid climate.

Throughout much of the territory there are only two distinct seasons—winter and summer; spring and autumn are usually brief periods of change between extremely low temperatures and extremely high.[] The coldest month is January (February on the coastline), the warmest usually is July. Great ranges of temperature are typical. In winter, temperatures get colder both from south to north and from west to east. Summers can be quite hot, even in Siberia.[109] The continental interiors are the driest areas.

Biodiversity

The brown bear is a popular symbol of Russia, particularly in the West.

From north to south the East European Plain, also known as Russian Plain, is clad sequentially in Arctic tundra, coniferous forest (taiga), mixed and broad-leaf forests, grassland (steppe), and semi-desert (fringing the Caspian Sea), as the changes in vegetation reflect the changes in climate. Siberia supports a similar sequence but is largely taiga. Russia has the world's largest forest reserves,[] known as "the lungs of Europe",[110] second only to the Amazon Rainforest in the amount of carbon dioxide it absorbs.

There are 266 mammal species and 780 bird species in Russia. A total of 415 animal species have been included in the Red Data Book of the Russian Federation as of 1997[111] and are now protected.

Economy

World Trade Center in Moscow

Russia has a market economy with enormous natural resources, particularly oil and natural gas. It has the 8th largest economy in the world by nominal GDP and the 6th largest by purchasing power parity (PPP). Since the turn of the 21st century, higher domestic consumption and greater political stability have bolstered economic growth in Russia. The country ended 2008 with its ninth straight year of growth, averaging 7% annually between 2000 and 2008. Real GDP per capita, PPP (current international $) was 19,840 in 2010.[] Growth was primarily driven by non-traded services and goods for the domestic market, as opposed to oil or mineral extraction and exports.[17] The average nominal salary in Russia was $640 per month in early 2008, up from $80 in 2000.[112] In the end of 2011 the average nominal monthly wages reached 24,310 RUR (or US$810),[113] while tax on the income of individuals is payable at the rate of 13% on most incomes.[114] Approximately 13.7% of Russians lived below the national poverty line in 2010,[115] significantly down from 40% in 1998 at the worst point of the post-Soviet collapse.[] Unemployment in Russia was at 6% in 2007, down from about 12.4% in 1999.[116] The middle class has grown from just 8 million persons in 2000 to 55 million persons in 2006.[117] Sugar imports reportedly dropped 82% between 2012–2013.[118]

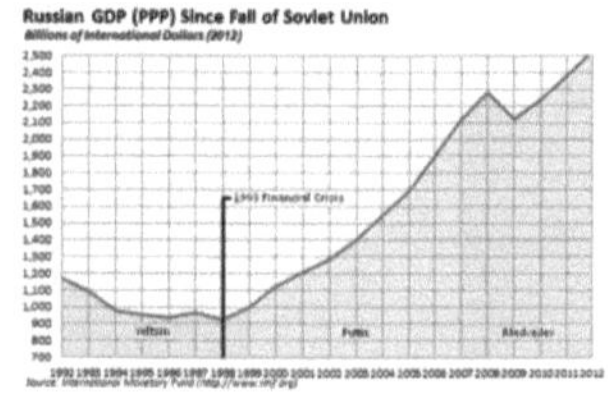

Russian economy since the end of the Soviet Union

Oil, natural gas, metals, and timber account for more than 80% of Russian exports abroad.[17] Since 2003, the exports of natural resources started decreasing in economic importance as the internal market strengthened considerably. Despite higher energy prices, oil and gas only contribute to 5.7% of Russia's GDP and the government predicts this will be 3.7% by 2011.[119] Oil export earnings allowed Russia to increase its foreign reserves from $12 billion in 1999 to $597.3 billion on 1 August 2008, the third largest foreign exchange reserves in the world.[120] The macroeconomic policy under Finance Minister Alexei Kudrin was prudent and sound, with excess income being stored in the

Stabilization Fund of Russia.[] In 2006, Russia repaid most of its formerly massive debts,[121] leaving it with one of the lowest foreign debts among major economies.[122] The Stabilization Fund helped Russia to come out of the global financial crisis in a much better state than many experts had expected.[]

A simpler, more streamlined tax code adopted in 2001 reduced the tax burden on people and dramatically increased state revenue.[123] Russia has a flat tax rate of 13%. This ranks it as the country with the second most attractive personal tax system for single managers in the world after the United Arab Emirates.[124] According to Bloomberg, Russia is considered well ahead of most other resource-rich countries in its economic development, with a long tradition of education, science, and industry.[125] The country has a higher proportion of higher education graduates than any other country in Eurasia.[126]

The economic development of the country has been uneven geographically with the Moscow region contributing a very large share of the country's GDP.[127] Another problem is modernisation of infrastructure, ageing and inadequate after years of being neglected in 1990s; the government has said $1 trillion will be invested in development of infrastructure by 2020.[128] In December 2011, Russia finally joined World Trade Organisation, allowing it a greater access to overseas markets. Some analysts estimate that WTO membership could bring the Russian economy a bounce of up to 3 per cent annually.[129] Russia ranks the second most corrupt country in Europe (after Ukraine), according to the Corruption Perceptions Index. The Norwegian-Russian Chamber of Commerce also states that *"[c]orruption is one of the biggest problems both Russian and international companies have to deal with."*[130]

Agriculture

The total area of cultivated land in Russia was estimated as 1,237,294 km^2 in 2005, the fourth largest in the world.[131] In 1999–2009, Russia's agriculture demonstrated steady growth,[132] and the country turned from a grain importer to the third largest grain exporter after EU and USA.[133] The production of meat has grown from 6,813,000 tonnes in 1999 to 9,331,000 tonnes in 2008, and continues to grow.[134]

Rye Fields, by Ivan Shishkin. Russia is the world's top producer of rye, barley, buckwheat, oats and sunflower seed, and one of the largest producers and exporters of wheat.

This restoration of agriculture was supported by credit policy of the government, helping both individual farmers and large privatized corporate farms, that once were Soviet kolkhozes and still own the significant share of agricultural land.[135] While large farms concentrate mainly on the production of grain and husbandry products, small private household plots produce most of the country's yield of potatoes, vegetables and fruits.[136]

With access to three of the world's oceans—the Atlantic, Arctic, and Pacific—Russian fishing fleets are a major contributor to the world's fish supply. The total capture of fish was at 3,191,068 tons in 2005.[137] Both exports and imports of fish and sea products grew significantly in the recent years, reaching correspondingly $2,415 and $2,036 millions in 2008.[138]

Sprawling from the Baltic Sea to the Pacific Ocean, Russia has more than a fifth of the world's forests, which makes it the largest forest country in the world.[] [139] However, according to a 2012 study by the Food and Agriculture Organization of the United Nations and the Government of the Russian Federation,[140] the considerable potential of Russian forests is underutilized and Russia's share of the global trade in forest products is less than four percent.[141] [142]

Energy

Russia is a key oil and gas supplier to much of Europe.

In recent years, Russia has frequently been described in the media as an energy superpower.[143] [144] The country has the world's largest natural gas reserves,[145] the 8th largest oil reserves,[146] and the second largest coal reserves.[147] Russia is the world's leading natural gas exporter[148] and second largest natural gas producer,[17] while also the largest oil exporter and the largest oil producer.[16] On 1 January 2011, Russia said it had begun scheduled oil shipments to China, with the plan to increase the rate up to 300,000 barrels per day in 2011.[149]

Russia is the 3rd largest electricity producer in the world[150] and the 5th largest renewable energy producer, the latter because of the well-developed hydroelectricity production in the country.[151] Large cascades of hydropower plants are built in European Russia along big rivers like Volga. The Asian part of Russia also features a number of major hydropower stations, however the gigantic hydroelectric potential of Siberia and the Russian Far East largely remains unexploited.

Russia was the first country to develop civilian nuclear power and to construct the world's first nuclear power plant. Currently the country is the 4th largest nuclear energy producer,[152] with all nuclear power in Russia being managed by Rosatom State Corporation. The sector is rapidly developing, with an aim of increasing the total share of nuclear energy from current 16.9% to 23% by 2020. The Russian government plans to allocate 127 billion rubles ($5.42 billion) to a federal program dedicated to the next generation of nuclear energy technology. About 1 trillion rubles ($42.7 billion) is to be allocated from the federal budget to nuclear power and industry development before 2015.[153]

Transport

Railway transport in Russia is mostly under the control of the state-run Russian Railways monopoly. The company accounts for over 3.6% of Russia's GDP and handles 39% of the total freight traffic (including pipelines) and more than 42% of passenger traffic.[] The total length of common-used railway tracks exceeds 85,500 km,[] second only to the U.S. Over 44,000 km of tracks are electrified,[154] which is the largest number in the world, and additionally there are more than 30,000 km of industrial non-common carrier lines. Railways in Russia, unlike in the most of the world, use broad gauge of , with the exception of 957 km on Sakhalin island using narrow gauge of . The most renown railway in Russia is Trans-Siberian (*Transsib*), spanning a record 7 time zones and serving the longest single continuous services in the world, Moscow-Vladivostok (9,259 km, 5,753 mi), Moscow–Pyongyang (10,267 km, 6,380 mi)[155] and Kiev–Vladivostok (11,085 km, 6,888 mi).[156]

The marker for kilometre 9288, at the end of the Trans-Siberian Railway in Vladivostok

As of 2006 Russia had 933,000 km of roads, of which 755,000 were paved.[157] Some of these make up the Russian federal motorway system. With a large land area the road density is the lowest of all the G8 and BRIC countries.[158]

Much of Russia's inland waterways, which total 102000 km (mi), are made up of natural rivers or lakes. In the European part of the country the network of channels connects the basins of major rivers. Russia's capital, Moscow, is sometimes called *"the port of the five seas"*, because of its waterway connections to the Baltic, White, Caspian, Azov and Black Seas.

Major sea ports of Russia include Rostov-on-Don on the Azov Sea, Novorossiysk on the Black Sea, Astrakhan and Makhachkala on the Caspian, Kaliningrad and St Petersburg on the Baltic, Arkhangelsk on the White Sea, Murmansk on the Barents Sea, Petropavlovsk-Kamchatsky and Vladivostok on the Pacific Ocean. In 2008 the country owned 1448 merchant marine ships. The world's only fleet of nuclear-powered icebreakers advances the economic exploitation of the Arctic continental shelf of Russia and the development of sea trade through the Northern Sea Route between Europe and East Asia.

Yamal, one of Russia's nuclear-powered icebreakers (gallery [159])

By total length of pipelines Russia is second only to the U.S. Currently many new pipeline projects are being realized, including Nord Stream and South Stream natural gas pipelines to Europe, and the Eastern Siberia – Pacific Ocean oil pipeline (ESPO) to the Russian Far East and China.

Russia has 1216 airports,[160] the busiest being Sheremetyevo, Domodedovo, and Vnukovo in Moscow, and Pulkovo in St Petersburg. The total length of runways in Russia exceeds **unknown operator: u','** kilometres (**unknown operator: u'strong'** mi).[161]

Typically, major Russian cities have well-developed systems of public transport, with the most common varieties of exploited vehicles being bus, trolleybus and tram. Seven Russian cities, namely Moscow, Saint Petersburg, Nizhny Novgorod, Novosibirsk, Samara, Yekaterinburg and Kazan, have underground metros, while Volgograd features a metrotram. The total length of metros in Russia is 465.4 kilometres (mi). Moscow Metro and Saint Petersburg Metro are the oldest in Russia, opened in 1935 and 1955 respectively. These two are among the fastest and busiest metro systems in the world, and are famous for rich decorations and unique designs of their stations, which is a common tradition on Russian metros and railways.

Science and technology

Science and technology in Russia blossomed since the Age of Enlightenment, when Peter the Great founded the Russian Academy of Sciences and Saint Petersburg State University, and polymath Mikhail Lomonosov established the Moscow State University, paving the way for a strong native tradition in learning and innovation. In the 19th and 20th centuries the country produced a large number of notable scientists and inventors.

Mikhail Lomonosov, polymath scientist, inventor, poet and artist

The Russian physics school began with Lomonosov who proposed the law of conservation of matter preceding the energy conservation law. Russian discoveries and inventions in physics include the electric arc, electrodynamical Lenz's law, space groups of crystals, photoelectric cell, Cherenkov radiation, electron paramagnetic resonance, heterotransistors and 3D holography. Lasers and masers were co-invented by Nikolai Basov and Alexander Prokhorov, while the idea of tokamak for controlled nuclear fusion was introduced by Igor Tamm, Andrei Sakharov and Lev Artsimovich, leading eventually the modern international ITER project, where Russia is a party.

Since the time of Nikolay Lobachevsky (a *Copernicus of Geometry* who pioneered the non-Euclidean geometry) and a prominent tutor Pafnuty Chebyshev, the Russian mathematical school became one of the most influential in the world.[162] Chebyshev's students included Aleksandr Lyapunov, who founded the modern stability theory, and

Andrey Markov who invented the Markov chains. In the 20th century Soviet mathematicians, such as Andrey Kolmogorov, Israel Gelfand and Sergey Sobolev, made major contributions to various areas of mathematics. Nine Soviet/Russian mathematicians were awarded with Fields Medal, a most prestigious award in mathematics. Recently Grigori Perelman was offered the first ever Clay Millennium Prize Problems Award for his final proof of the Poincaré conjecture in 2002.[163]

Russian chemist Dmitry Mendeleev invented the Periodic table, the main framework of modern chemistry. Aleksandr Butlerov was one of the creators of the theory of chemical structure, playing a central role in organic chemistry. Russian biologists include Dmitry Ivanovsky who discovered viruses, Ivan Pavlov who was the first to experiment with the classical conditioning, and Ilya Mechnikov who was a pioneer researcher of the immune system and probiotics.

Soyuz TMA-2 is launched from Baikonur, Kazakhstan, carrying one of the first resident crews to the International Space Station

Many Russian scientists and inventors were émigrés, like Igor Sikorsky, who built the first airliners and modern-type helicopters; Vladimir Zworykin, often called the father of TV; chemist Ilya Prigogine, noted for his work on dissipative structures and complex systems; Nobel Prize-winning economists Simon Kuznets and Wassily Leontief; physicist Georgiy Gamov (an author of the Big Bang theory) and social scientist Pitirim Sorokin. Many foreigners worked in Russia for a long time, like Leonard Euler and Alfred Nobel.

Russian inventions include the arc welding by Nikolay Benardos, further developed by Nikolay Slavyanov, Konstantin Khrenov and other Russian engineers. Gleb Kotelnikov invented the knapsack parachute, while Evgeniy Chertovsky introduced the pressure suit. Alexander Lodygin and Pavel Yablochkov were pioneers of electric lighting, and Mikhail Dolivo-Dobrovolsky introduced the first three-phase electric power systems, widely used today. Sergei Lebedev invented the first commercially viable and mass-produced type of synthetic rubber. The first ternary computer, *Setun*, was developed by Nikolay Brusentsov.

In the 20th century a number of prominent Soviet aerospace engineers, inspired by the fundamental works of Nikolai Zhukovsky, Sergei Chaplygin and others, designed many hundreds of models of military and civilian aircraft and founded a number of *KBs* (*Construction Bureaus*) that now constitute the bulk of Russian United Aircraft Corporation. Famous Russian aircraft include the civilian Tu-series, Su and MiG fighter aircraft, Ka and Mi-series helicopters; many Russian aircraft models are on the list of most produced aircraft in history.

Famous Russian battle tanks include T-34, the best tank design of World War II,[164] and further tanks of T-series, including the most produced tank in history, T-54/55.[165] The AK-47 and AK-74 by Mikhail Kalashnikov constitute the most widely used type of assault rifle throughout the world—so much so that more AK-type rifles have been manufactured than all other assault rifles combined.[166]

The Sukhoi PAK FA is a fifth-generation jet fighter being developed for the Russian Air Force.

Yo-mobil, a new Russian hybrid car

With all these achievements, however, since the late Soviet era Russia was lagging behind the West in a number of technologies, mostly those related to energy conservation and consumer goods production. The crisis of 1990-s led to the drastic reduction of the state support for science and a brain drain migration from Russia.

In the 2000s, on the wave of a new economic boom, the situation in the Russian science and technology has improved, and the government launched a campaign aimed into modernisation and innovation. Russian President Dmitry Medvedev formulated top 5 priorities for the country's technological development: efficient energy use, IT (including both common products and the products combined with space technology), nuclear energy and pharmaceuticals.[167]

Currently Russia has completed the GLONASS satellite navigation system. The country is developing its own fifth-generation jet fighter and constructing the first serial mobile nuclear plant in the world. In 2010, an economy class hybrid electric car project was introduced, called Yo-mobil, that will be mass-produced by ë-Auto, a Russian company that is a joint venture between truck maker Yarovit and the Onexim investment group.[] [] []

Space exploration

Soviet and Russian space station Mir

Russian achievements in the field of space technology and space exploration are traced back to Konstantin Tsiolkovsky, the father of theoretical austronautics.[168] His works had inspired leading Soviet rocket engineers, such as Sergey Korolyov, Valentin Glushko and many others who contributed to the success of the Soviet space program on early stages of the Space Race and beyond.

In 1957 the first Earth-orbiting artificial satellite, *Sputnik 1*, was launched; in 1961 the first human trip into space was successfully made by Yury Gagarin. Many other Soviet and Russian space exploration records ensued, including the first spacewalk performed by Alexey Leonov, Luna 9 was the first spacecraft to land on the Moon, Venera 7 was the first to land on another planet (Venus), Mars 3 then the first to land on Mars, the first space exploration rover *Lunokhod-1* and the first space station *Salyut 1* and *Mir*.

After the collapse of the Soviet Union, some government-funded space exploration programs, including the Buran space shuttle program, were cancelled or delayed, while participation of the Russian space industry in commercial activities and international cooperation intensified. Nowadays Russia is the largest satellite launcher.[169] After the U.S. Space Shuttle program ended in 2011, Soyuz rockets became the only provider of transport for astronauts at the International Space Station.

Demographics

Percentage of ethnic Russians by region in 2010

>80%

70—79%

50—69%

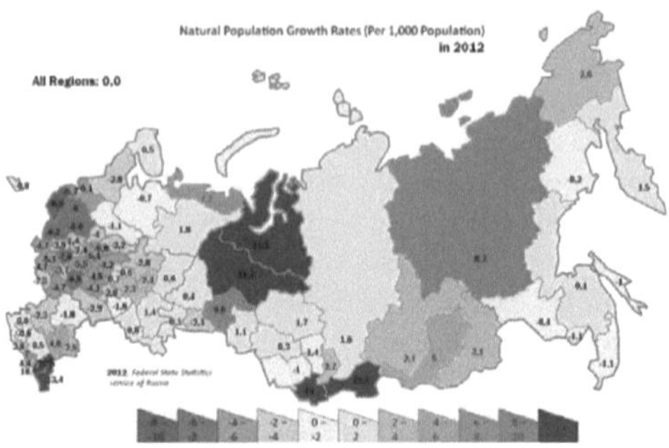

Natural population growth rate in Russia, 2012.

Ethnic Russians comprise 81% of the country's population.[170] The Russian Federation is also home to several sizeable minorities. In all, 160 different other ethnic groups and indigenous peoples live within its borders.[171] Though Russia's population is comparatively large, its density is low because of the country's enormous size. Population is densest in European Russia, near the Ural Mountains, and in southwest Siberia. 73% of the population lives in urban areas while 27% in rural ones.[172] The results of the 2010 Census show a total population of 142,856,536.[]

Russia's population peaked at 148,689,000 in 1991, just before the dissolution of the Soviet Union. It began to experience a rapid decline starting in the mid-1990s.[173] The decline has slowed to near stagnation in recent years because of reduced death rates, increased birth rates and increased immigration.[174]

In 2009, Russia recorded annual population growth for the first time in fifteen years, with total growth of 10,500.[174] 279,906 migrants arrived to the Russian Federation the same year, of which 93% came from CIS countries.[174] The number of Russian emigrants steadily declined from 359,000 in 2000 to 32,000 in 2009.[174] There are also an estimated 10 million illegal immigrants from the ex-Soviet states in Russia.[175] Roughly 116 million ethnic Russians live in Russia[171] and about 20 million more live in other former republics of the Soviet Union,[176] mostly in Ukraine and Kazakhstan.[177]

The 2010 census recorded 81% of the population as ethnically Russian, and 19% as other ethnicities:[170] 3.7% Tatars; 1.4% Ukrainians; 1.1% Bashkirs; 1% Chuvashes; 11.8% others and unspecified. According to the Census,

84.93% of the Russian population belongs to European ethnic groups (Slavic, Germanic, Finnic other than Ugric, Greek.etc.). This is a decline from the 2002, when they constituted for more than 86% of the population.[170]

Russia's birth rate is higher than that of most European countries (12.6 births per 1000 people in 2010[174] compared to the European Union average of 9.90 per 1000),[178] but its death rate is also substantially higher (in 2010, Russia's death rate was 14.3 per 1000 people[174] compared to the EU average of 10.28 per 1000).[179] The Russian Ministry of Health and Social Affairs predicted that by 2011 the death rate would equal the birth rate because of increase in fertility and decline in mortality.[180] The government is implementing a number of programs designed to increase the birth rate and attract more migrants. Monthly government child assistance payments were doubled to US$55, and a one-time payment of US$9,200 was offered to women who had a second child since 2007.[181] In 2009 Russia experienced its highest birth rate since the dissolution of the Soviet Union.[174] [182]

Language

Russia's 160 ethnic groups speak some 100 languages.[] According to the 2002 Census, 142.6 million people speak Russian, followed by Tatar with 5.3 million and Ukrainian with 1.8 million speakers.[183] Russian is the only official state language, but the Constitution gives the individual republics the right to make their native language official in addition to Russian.[184]

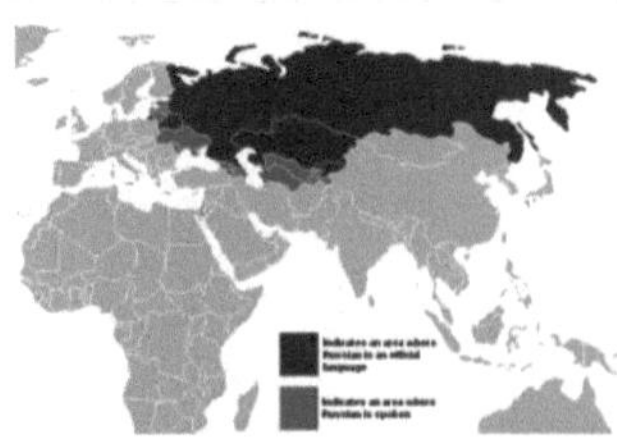
Countries where the Russian language is spoken

Despite its wide distribution, the Russian language is homogeneous throughout the country. Russian is the most geographically widespread language of Eurasia, as well as the most widely spoken Slavic language.[185] It belongs to the Indo-European language family and is one of the living members of the East Slavic languages, the others being Belarusian and Ukrainian (and possibly Rusyn). Written examples of Old East Slavic (*Old Russian*) are attested from the 10th century onwards.[186]

Russian is one of the six official languages of the UN.[187]

Religion

The Temple of All Religions in Kazan

Orthodox Christianity, Islam, Judaism and Buddhism are Russia's traditional religions, and are all legally a part of Russia's "historical heritage".[188] Traced back to the Christianization of Kievan Rus' in the 10th century, Russian Orthodoxy is the dominant religion in the country; smaller Christian denominations such as Catholics, Armenian Gregorians and various Protestant churches also exist.. The Russian Orthodox Church was the country's state religion prior to the Revolution and remains the largest religious body in the country. Estimates on how many believers the church has fluctuate widely in different sources. A 15 January 2013 statistic by the Russian Academy of Sciences announced that 79% of the Russian population belong to the Russian Orthodox Church.[] 95% of the registered Orthodox parishes belong to the Russian Orthodox Church while there are a number of smaller Orthodox Churches.[189] However, the vast majority of Orthodox believers do not attend church on a regular basis. Easter is the most popular religious holiday in Russia, celebrated by more than 90% of all Russian citizens, including large numbers of those who are non-religious. More than three-quarters of the Russian population celebrate Easter by making traditional Easter cakes, coloured eggs and paskha.[190]

Estimates of the number of Muslims in Russia range from 7–9 million by local sources, to 15–20 million by Western and Islamic sources.[191] There are approximately 3 to 4 million temporary Muslim migrants from the post-Soviet states.[192] Most Muslims live in the Volga-Ural region, as well as in the Caucasus, Moscow, St. Petersburg and Western Siberia.[193]

Buddhism is traditional in three regions of the Russian Federation: Buryatia, Tuva, and Kalmykia. Some residents of the Siberian and Far Eastern regions, such as Yakutia and Chukotka, practice shamanist, pantheistic, and pagan rites, along with the major religions. Induction into religion takes place primarily along ethnic lines. Slavs are overwhelmingly Orthodox Christian, Turkic speakers are predominantly Muslim, and Mongolic peoples are generally Buddhists.[194]

Various reports put the number of non-religious in Russia at between 16–48% of the population.[195] The number of atheists has decreased significantly; according to the recent statistic, only seven percent declared themselves atheists, a decrease of 5% in three years.[]

Health

A mobile clinic used to provide health care at remote railway stations

The Russian Constitution guarantees free, universal health care for all its citizens.[196] In practice, however, free health care is partially restricted because of mandatory registration.[197] While Russia has more physicians, hospitals, and health care workers than almost any other country in the world on a per capita basis,[198] since the dissolution of the Soviet Union the health of the Russian population has declined considerably as a result of social, economic, and lifestyle changes;[199] the trend has been reversed only in the recent years, with average life expectancy having increased 2.4 years for males and 1.4 years for females between 2006–09.[174]

As of 2009, the average life expectancy in Russia was 62.77 years for males and 74.67 years for females.[200] The biggest factor contributing to the relatively low life expectancy for males is a high mortality rate among working-age males. Deaths mostly occur because of preventable causes (e.g., alcohol poisoning, smoking, traffic accidents, violent crime).[174] As a result of the large gender difference in life expectancy, and also because of the lasting effect of high casualties in World War II, the gender imbalance remains to this day and there are 0.859 males to every female.[17]

Education

A school in Moscow. The tower of the Moscow State University is seen at the distance.

Russia has a free education system – this is guaranteed for all citizens by the Constitution,[201] however an entry to subsidized post-secondary education is highly competitive.[202] As a result of great emphasis on science and technology in education, Russian medical, mathematical, scientific, and aerospace research is generally of a high order.[203]

Since 1990, the 11-year school education has been introduced. Education in state-owned secondary schools is free. University level education is free, with exceptions. A substantial share of students is enrolled for full pay (many state institutions started to open commercial positions in the last years).[204]

In 2004, state spending for education amounted to 3.6% of the GDP, or 13% of the consolidated state budget.[205] The Government allocates funding to pay the tuition fees within an established quota or number of students for each state institution. In higher education institutions, students are paid a small stipend and provided with free housing if they are from out of town.[206]

The oldest and largest Russian universities are Moscow State University and Saint Petersburg State University. In the 2000s, in order to create higher education and research institutions of comparable scale in Russian regions, the government launched a program of establishing *federal universities*, mostly by merging existing large regional universities and research institutes and providing them with a special funding. These new institutions include the Southern Federal University, Siberian Federal University, Kazan Volga Federal University, North-Eastern Federal University and Far Eastern Federal University.

Culture

Folk culture and cuisine

There are over 160 different ethnic groups and indigenous peoples in Russia.[171] Ethnic Russians with their Slavic Orthodox traditions, Tatars and Bashkirs with their Turkic Muslim culture, Buddhist nomadic Buryats and Kalmyks, Shamanistic peoples of the Extreme North and Siberia, highlanders of the Northern Caucasus, Finno-Ugric peoples of the Russian North West and Volga Region all contribute to the cultural diversity of the country.

The Merchant's Wife by Boris Kustodiev, showcasing the Russian tea culture

Handicraft, like Dymkovo toy, khokhloma, gzhel and palekh miniature represent an important aspect of Russian folk culture. Ethnic Russian clothes include kaftan, kosovorotka and ushanka for men, sarafan and kokoshnik for women, with lapti and valenki as common shoes. The clothes of Cossacks from Southern Russia include burka and papaha, which they share with the peoples of the Northern Caucasus.

Russian cuisine widely uses fish, poultry, mushrooms, berries, and honey. Crops of rye, wheat, barley, and millet provide the ingredients for various breads, pancakes and cereals, as well as for kvass, beer and vodka drinks. Black bread is rather popular in Russia, compared to the rest of the world. Flavourful soups and stews include shchi, borsch, ukha, solyanka and okroshka. Smetana (a heavy sour cream) is often added to soups and salads. Pirozhki, blini and syrniki are native types of pancakes. Chicken Kiev, pelmeni and shashlyk are popular meat dishes, the last two being of Tatar and Caucasus origin respectively. Other meat dishes include stuffed cabbage rolls *(golubtsy)* usually filled with meat.[] Salads include Olivier salad, Vinaigrette and Dressed herring.

Russia's large number of ethnic groups have distinctive traditions regarding folk music. Typical ethnic Russian musical instruments are gusli, balalaika, zhaleika and garmoshka. Folk music had a significant influence on Russian classical composers, and in modern times it is a source of inspiration for a number of popular folk bands, like Melnitsa. Russian folk songs, as well as patriotic Soviet songs, constitute the bulk of the repertoire of the world-renown Red Army choir and other popular ensembles.

Russians have many traditions, including the washing in banya, a hot steam bath somewhat similar to sauna.[38] Old Russian folklore takes its roots in the pagan Slavic religion. Many Russian fairy tales and epic bylinas were adaptated for animation films, or for feature movies by the prominent directors like Aleksandr Ptushko (*Ilya Muromets*, *Sadko*) and Aleksandr Rou (*Morozko*, *Vasilisa the Beautiful*). Russian poets, including Pyotr Yershov and Leonid Filatov, made a number of well-known poetical interpretations of the classical fairy tales, and in some cases, like that of Alexander Pushkin, also created fully original fairy tale poems of great popularity.

Architecture

Saint Basil's Cathedral in Moscow, a well known piece of Russian architecture

Since the Christianization of Kievan Rus' for several ages Russian architecture was influenced predominantly by the Byzantine architecture. Apart from fortifications (kremlins), the main stone buildings of ancient Rus' were Orthodox churches with their many domes, often gilded or brightly painted.

Aristotle Fioravanti and other Italian architects brought Renaissance trends into Russia since the late 15th century, while the 16th century saw the development of unique tent-like churches[207] culminating in Saint Basil's Cathedral. By that time the onion dome design was also fully developed.[208] In the 17th century, the "fiery style" of ornamentation flourished in Moscow and Yaroslavl, gradually paving the way for the Naryshkin baroque of the 1690s. After the reforms of Peter the Great the change of architectural styles in Russia generally followed that in the Western Europe.

The 18th-century taste for rococo architecture led to the ornate works of Bartolomeo Rastrelli and his followers. The reigns of Catherine the Great and her grandson Alexander I saw the flourishing of Neoclassical architecture, most notably in the capital city of Saint Petersburg. The second half of the 19th century was dominated by the Neo-Byzantine and Russian Revival styles. Prevalent styles of the 20th century were the Art Nouveau, Constructivism, and the Stalin Empire style.

In 1955, a new Soviet leader, Nikita Khrushchev, condemned the "excesses" of the former academic architecture,[209] and the late Soviet era was dominated by plain functionalism in architecture. This helped somewhat to resolve the housing problem, but created a large quantity of buildings of low architectural quality, much in contrast with the previous bright styles. The situation improved in the recent two decades. Many temples demolished in Soviet times were rebuilt, and this process continues along with the restoration of various historical buildings destroyed in World War II. A total of 23,000 Orthodox churches have been rebuilt between 1991–2010, which effectively quadrapled the number of operating churches in Russia.[210] For many, mostly elderly and conservative inhabitants, the lost Russia is identified with the Soviets. For a more liberal and largely middle-aged part of the Russian population it is the pre-revolutionary empire.[]

Visual arts

The *Trinity* icon by Andrei Rublev

Early Russian painting is represented in icons and vibrant frescos, the two genres inherited from Byzantium. As Moscow rose to power, Theophanes the Greek, Dionisius and Andrei Rublev became vital names associated with a distinctly Russian art.

The Russian Academy of Arts was created in 1757[211] and gave Russian artists an international role and status. Ivan Argunov, Dmitry Levitzky, Vladimir Borovikovsky and other 18th century academicians mostly focused on portrait painting. In the early 19th century, when neoclassicism and romantism flourished, mythological and Biblical themes inspired many prominent painings, notably by Karl Briullov and Alexander Ivanov.

In the mid-19th century the *Peredvizhniki* (*Wanderers*) group of artists broke with the Academy and initiated a school of art liberated from academic restrictions.[212] These were mostly realist painters who captured Russian identity in landscapes of wide rivers, forests, and birch clearings, as well as vigorous genre scenes and robust portraits of their contemporaries. Some artists focused on depicting dramatic moments in Russian history, while others turned to social criticism, showing the conditions of the poor and caricaturing authority; critical realism flourished under the reign of Alexander II. Leading realists include Ivan Shishkin, Arkhip Kuindzhi, Ivan Kramskoi, Vasily Polenov, Isaac Levitan, Vasily Surikov, Viktor Vasnetsov, Ilya Repin and Boris Kustodiev.

The turn of the 20th century saw the rise of symbolist painting, represented by Mikhail Vrubel, Kuzma Petrov-Vodkin and Nicholas Roerich.

The Russian avant-garde was a large, influential wave of modernist art that flourished in Russia from approximately 1890 to 1930. The term covers many separate, but inextricably related art movements that occurred at the time, namely neo-primitivism, suprematism, constructivism, rayonism, and Russian Futurism. Notable artists from this era include El Lissitzky, Kazimir Malevich, Wassily Kandinsky, and Marc Chagall. Since 1930s the revolutionary ideas of the avant-garde clashed with the newly emerged conservative direction of socialist realism.

Soviet art produced works that were furiously patriotic and anti-fascist during and after the Great Patriotic War. Multiple war memorials, marked by a great restrained solemnity, were built throughout the country. Soviet artists often combined innovation with socialist realism, notably the sculptors Vera Mukhina, Yevgeny Vuchetich and Ernst Neizvestny.

Music and dance

The *Snowdance* scene from *The Nutcracker* ballet, composed by Pyotr Ilyich Tchaikovsky

Music in 19th century Russia was defined by the tension between classical composer Mikhail Glinka along with other members of The Mighty Handful, who embraced Russian national identity and added religious and folk elements to their compositions, and the Russian Musical Society led by composers Anton and Nikolay Rubinsteins, which was musically conservative. The later tradition of Pyotr Ilyich Tchaikovsky, one of the greatest composers of the Romantic era, was continued into the 20th century by Sergei Rachmaninoff.[213] World-renowned composers of the 20th century include Alexander Scriabin, Igor Stravinsky, Sergei Prokofiev, Dmitri Shostakovich and Alfred Schnittke.

Russian conservatories have turned out generations of famous soloists. Among the best known are violinists David Oistrakh and Gidon Kremer; cellist Mstislav Rostropovich; pianists Vladimir Horowitz, Sviatoslav Richter, and Emil Gilels; and vocalists Fyodor Shalyapin, Galina Vishnevskaya, Anna Netrebko and Dmitry Hvorostovsky.[214]

During the early 20th century, Russian ballet dancers Anna Pavlova and Vaslav Nijinsky rose to fame, and impresario Sergei Diaghilev and his Ballets Russes' travels abroad profoundly influenced the development of dance worldwide.[215] Soviet ballet preserved the perfected 19th century traditions,[216] and the Soviet Union's choreography schools produced many internationally famous stars, including Maya Plisetskaya, Rudolf Nureyev, and Mikhail Baryshnikov. The Bolshoi Ballet in Moscow and the Mariinsky Ballet in St Petersburg remain famous throughout the world.[217]

Modern Russian rock music takes its roots both in the Western rock and roll and heavy metal, and in traditions of the Russian bards of the Soviet era, such as Vladimir Vysotsky and Bulat Okudzhava.[218] Popular Russian rock groups include Mashina Vremeni, DDT, Aquarium, Alisa, Kino, Kipelov, Nautilus Pompilius, Aria, Grazhdanskaya Oborona, Splean and Korol i Shut. Russian pop music developed from what was known in the Soviet times as *estrada* into full-fledged industry, with some performers gaining wide international recognition, such as t.A.T.u., Nu Virgos and Vitas.

Literature and philosophy

Leo Tolstoy, novelist and philosopher

In the 18th century, during the era of Russian Enlightenment, the development of Russian literature was boosted by the works of Mikhail Lomonosov and Denis Fonvizin. By the early 19th century a modern native tradition had emerged, producing some of the greatest writers in Russian history. This period, known also as the Golden Age of Russian Poetry, began with Alexander Pushkin, who is considered the founder of the modern Russian literary language and often described as the "Russian Shakespeare".[219] It continued into the 19th century with the poetry of Mikhail Lermontov and Nikolay Nekrasov, dramas of Alexander Ostrovsky and Anton Chekhov, and the prose of Nikolai Gogol and Ivan Turgenev. Leo Tolstoy and Fyodor Dostoyevsky have been described by literary critics as the greatest novelists of all time.[220] [221]

By the 1880s, the age of the great novelists was over, and short fiction and poetry became the dominant genres. The next several decades became known as the Silver Age of Russian Poetry, when the previously dominant literary realism was replaced by symbolism. Leading authors of this era include such poets as Valery Bryusov, Vyacheslav Ivanov, Alexander Blok, Nikolay Gumilev and Anna Akhmatova, and novelists Leonid Andreyev, Ivan Bunin, and Maxim Gorky.

Russian philosophy blossomed in the 19th century, when it was defined initially by the opposition of Westernizers, advocating Western political and economical models, and Slavophiles, insisting on developing Russia as a unique civilization. The latter group includes Nikolai Danilevsky and Konstantin Leontiev, the founders of eurasianism. In its further development Russian philosophy was always marked by a deep connection to literature and interest in creativity, society, politics and nationalism; Russian cosmism and religious philosophy were other major areas. Notable philosophers of the late 19th and the early 20th centuries include Vladimir Solovyev, Sergei Bulgakov, and Vladimir Vernadsky.

Following the Russian Revolution of 1917 many prominent writers and philosophers left the country, including Bunin, Vladimir Nabokov and Nikolay Berdyayev, while a new generation of talented authors joined together in an effort to create a distinctive working-class culture appropriate for the new Soviet state. In the 1930s censorship over literature was tightened in line with the policy of socialist realism. In the late 1950s restrictions on literature were eased, and by the 1970s and 1980s, writers were increasingly ignoring official guidelines. Leading authors of the Soviet era include novelists Yevgeny Zamyatin, Ilf and Petrov, Mikhail Bulgakov and Mikhail Sholokhov, and poets Vladimir Mayakovsky, Yevgeny Yevtushenko, and Andrey Voznesensky.

The Soviet Union was also a major producer of science fiction, written by authors like Arkady and Boris Strugatsky, Kir Bulychov, Alexander Belayev and Ivan Yefremov.[222] Traditions of Russian science fiction and fantasy are continued today by numerous writers.

Cinema, animation and media

Former Russian President Dmitry Medvedev in the Washington studio of Russia Today TV

Russian and later Soviet cinema was a hotbed of invention in the period immediately following the 1917, resulting in world-renowned films such as *The Battleship Potemkin* by Sergei Eisenstein.[] Eisenstein was a student of filmmaker and theorist Lev Kuleshov, who developed the Soviet montage theory of film editing at the world's first film school, the All-Union Institute of Cinematography. Dziga Vertov, whose *kino-glaz* ("film-eye") theory—that the camera, like the human eye, is best used to explore real life—had a huge impact on the development of documentary film making and cinema realism. The subsequent state policy of socialist realism somewhat limited creativity, however many Soviet films in this style were artistically successful, like *Chapaev*, *The Cranes Are Flying*, and *Ballad of a Soldier*.[]

1960s and 1970s saw a greater variety of artistic styles in the Soviet cinema. Eldar Ryazanov's and Leonid Gaidai's comedies of that time were immensely popular, with many of the catch phrases still in use today. In 1961–68 Sergey Bondarchuk directed an Oscar-winning film adaptation of Leo Tolstoy's epic *War and Peace*, which was the most expensive film made in the Soviet Union.[223] In 1969, Vladimir Motyl's *White Sun of the Desert* was released, a very popular film in a genre of ostern; the film is traditionally watched by cosmonauts before any trip into space.[224]

Russian animation dates back to the late Russian Empire times. During Soviet era, Soyuzmultfilm studio was the largest animation producer. Soviet animators developed a great variety of pioneering techniques and aesthetic styles, with prominent directors including Ivan Ivanov-Vano, Fyodor Khitruk and Aleksandr Tatarsky. Many Soviet cartoon heroes, such as the Russian-style Winnie-the-Pooh, cute little Cheburashka, Wolf and Hare from *Nu, Pogodi!* are iconic images in Russia and many surrounding countries.

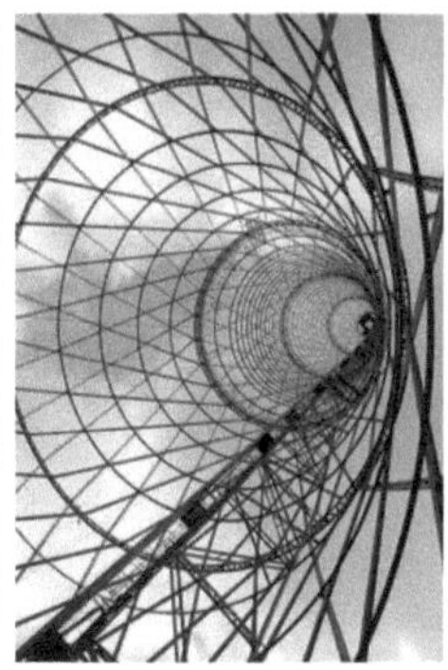

Shukhov Tower in Moscow served for the early radio and TV broadcasting.

The late 1980s and 1990s were a period of crisis in Russian cinema and animation. Although Russian filmmakers became free to express themselves, state subsidies were drastically reduced, resulting in fewer films produced. The early years of the 21st century have brought increased viewership and subsequent prosperity to the industry on the back of the economic revival. Production levels are already higher than in Britain and Germany.[225] Russia's total box-office revenue in 2007 was $565 million, up 37% from the previous year.[226] In 2002 the *Russian Ark* became the first feature film ever to be shot in a single take. The traditions of Soviet animation were developed recently by such directors as Aleksandr Petrov and studios like Melnitsa Animation.

Russia was among the first countries to introduce radio and television. While there were few channels in the Soviet time, in the past two decades many new state and private-owned radio stations and TV channels appeared. In 2005 a state-run English language Russia Today TV started broadcasting, and its Arabic version Rusiya Al-Yaum was launched in 2007.

Sports

Russian football fans with a gigantic "Go Russia!" banner, featuring the Russian Bear on the background of the Russian flag

Combining the total medals of Soviet Union and Russia, the country is second among all nations by number of gold medals both at the Summer Olympics and at the Winter Olympics. Soviet and later Russian athletes have always been in the top three for the number of gold medals collected at the Summer Olympics. Soviet gymnasts, track-and-field athletes, weight lifters, wrestlers, boxers, fencers, shooters, cross country skiers, biathletes, speed skaters and figure skaters were consistently among the best in the world, along with Soviet basketball, handball, volleyball and ice hockey players.[227] The 1980 Summer Olympics were held in Moscow while the 2014 Winter Olympics will be hosted in Sochi.

Although ice hockey was only introduced during the Soviet era, the national team managed to win gold at almost all the Olympics and World Championships they contested. Russian players Valery Kharlamov, Sergey Makarov, Vyacheslav Fetisov and Vladislav Tretiak hold four of six positions in the IIHF *Team of the Century*.[228] Recently Russia won the 2008, 2009 and the 2012 IIHF World Championships, overtaking Canada as the world's top ranked ice hockey team.[229] Russia totally dominated the 2012 tournament, winning all of its ten matches—the first time any team had done so since the Soviet Union in 1989.[230] The Kontinental Hockey League (KHL) was founded in 2008 as a successor to the Russian Superleague. It is seen as a rival to the National Hockey League (NHL) and is ranked the top hockey league in Europe as of 2009.[231] Bandy, also known as Russian hockey, is another traditionally popular ice sport.[232] The Soviet Union won all the Bandy World Championships between 1957–79.[233]

Dmitry Medvedev with Russia men's national ice hockey team

Along with ice hockey and basketball, association football is one of the most popular sports in modern Russia. The Soviet national team became the first ever European Champions by winning Euro 1960. In recent years, Russian football, which downgraded in 1990s, has experienced a revival. Russian clubs CSKA Moscow and Zenit St Petersburg won the UEFA Cup in 2005 and 2008 respectively. The Russian national football team reached the semi-finals of Euro 2008, losing only to the eventual champions Spain. Russia will host the 2018 FIFA World Cup, with 14 host cities located in the European part of the country and on the Urals. In 2007, the Russian national basketball team also won the European Basketball Championship. Russian basketball club PBC CSKA Moscow is one of the top teams in Europe, winning the Euroleague in 2006 and 2008.

Larisa Latynina, who currently holds a record for most Olympic medals won per person and most gold Olympic medals won by a woman, established the USSR as the dominant force in gymnastics for many years to come.[234] Today, Russia is leading in rhythmic gymnastics with Alina Kabayeva, Irina Tschaschina and Yevgeniya Kanayeva. Russian synchronized swimming is the best in the world, with almost all gold medals at Olympics and World Championships having been swept by Russians in recent decades. Figure skating is another popular sport in Russia, especially pair skating and ice dancing. At every Winter Olympics from 1964 until 2006 a Soviet or Russian pair has won gold. Since the end of the Soviet era, tennis has grown in popularity and Russia has produced a number of famous players, including Maria Sharapova, the world's highest paid female athlete.[235] In martial arts, Russia produced the sport Sambo and many renown fighters, like Fedor Emelianenko. Chess is a widely popular pastime in Russia; from 1927, Russian grandmasters have held the world chess championship almost continuously.[236]

Formula One is also becoming increasingly popular in Russia. Renault's Vitaly Petrov is the only Russian Formula One driver to date. There have only ever been two Russian Grands Prix (in 1913 and 1914), but it is set to return for 2014, in a six-year deal.[237]

National holidays and symbols

2005 Victory Day (9 May) parade on Moscow's Red Square

There are seven public holidays in Russia,[238] except those always celebrated on Sunday. Russian New Year traditions resemble those of the Western Christmas, with New Year Trees and gifts, and Ded Moroz (Father Frost) playing the same role as Santa Claus. Orthodox Christmas falls on 7 January, because Russian Orthodox Church still follows the Julian calendar and all Orthodox holidays are 13 days after Western ones. Another two major Christian holidays are Easter and Trinity Sunday. Kurban Bayram and Uraza Bayram are celebrated by Russian Muslims.

Further Russian public holidays include Defender of the Fatherland Day (23 February), which honors Russian men, especially those serving in the army; International Women's Day (8 March), which combines the traditions of Mother's Day and Valentine's Day; Spring and Labor Day (1 May); Victory Day (9 May); Russia Day (12 June); and Unity Day (4 November), commemorating the popular uprising which expelled the Polish occupation force from Moscow in 1612.

Victory Day is the second most popular holiday in Russia; it commemorates the victory over Nazism in the Great Patriotic War. A huge military parade, hosted by the President of Russia, is annually organised in Moscow on Red Square. Similar parades took place in all major Russian cities and cities with the status *Hero city* or *City of Military Glory*.

Popular non-public holidays include Old New Year (New Year according to Julian Calendar on 14 January), Tatiana Day (students holiday on 25 January), Maslenitsa (a pre-Christian spring holiday a week before the Great Lent), Cosmonautics Day (in tribute to the first human trip into space), Ivan Kupala Day (another pre-Christian holiday on 7 July) and Peter and Fevronia Day (taking place on 8 July and being the Russian analogue of Valentine's Day, which focuses, however, on the family love and fidelity).

Scarlet Sails celebration on the Neva river in Saint Petersburg

State symbols of Russia include the Byzantine double-headed eagle, combined with St. George of Moscow in the Russian coat of arms. The Russian flag dates from the late Tsardom of Russia period and has been widely used since the time of the Russian Empire. The Russian anthem shares its music with the Soviet Anthem, though not the lyrics. The imperial motto *God is with us* and the Soviet motto *Proletarians of all countries, unite!* are now obsolete and no new motto has replaced them. The hammer and sickle and the full Soviet coat of arms are still widely seen in Russian cities as a part of old architectural decorations. The Soviet Red Stars are also encountered, often on military equipment and war memorials. The Red Banner continues to be honored, especially the Banner of Victory of 1945.

The Matryoshka doll is a recognizable symbol of Russia, and the towers of Moscow Kremlin and Saint Basil's Cathedral in Moscow are main Russia's architectural icons. Cheburashka is a mascot of the Russian national Olympic team. St. Mary, St. Nicholas, St. Andrew, St. George, St. Alexander Nevsky, St. Sergius of Radonezh and St. Seraphim of Sarov are Russia's patron saints. Chamomile is the national flower, while birch the national tree. The Russian bear is an animal symbol and a national personification of Russia, though this image has a Western origin and Russians themselves have accepted it only fairly recently. The native Russian national personification is Mother Russia.

Tourism

Grand Cascade in Peterhof, a popular tourist destination in Saint Petersburg

Tourism in Russia has seen rapid growth since the late Soviet times, first domestic tourism and then international tourism, fueled by the rich cultural heritage and great natural variety of the country. Major tourist routes in Russia include a journey around the Golden Ring of ancient cities, cruises on the big rivers like the Volga, and long journeys on the famous Trans-Siberian Railway.

The most visited destinations in Russia are Moscow and Saint Petersburg, the current and the former capitals of the country. Recognized as World Cities, they feature such world-renown museums as Tretyakov Gallery and Hermitage, famous theaters like Bolshoi and Mariinsky, ornate churches like Saint Basil's Cathedral, Cathedral of Christ the Saviour, Saint Isaac's Cathedral and Church of the Savior on Blood, impressive fortifications like Moscow Kremlin and Peter and Paul Fortress, beautiful squares and streets like Red Square, Palace Square, Tverskaya Street and Nevsky Prospect. Rich palaces and parks are found in the former imperial residences in suburbs of Moscow (Kolomenskoye, Tsaritsyno) and St Petersburg (Peterhof, Strelna, Oranienbaum, Gatchina, Pavlovsk and Tsarskoye Selo). Moscow displays the Soviet architecture at its best, along with modern skyscrapers, while St Petersburg, nicknamed *Venice of the North*, boasts of its classical architecture, many rivers, channels and bridges.

Kazan, the capital of Tatarstan, shows a mix of Christian Russian and Muslim Tatar cultures. The city has registered a brand *The Third Capital of Russia*, though a number of other major cities compete for this status, including Novosibirsk, Yekaterinburg and Nizhny Novgorod.

Matryoshka doll taken apart

Typical Russian souvenirs include matryoshka doll and other handicrafts, samovars for water heating, ushanka and papaha warm hats, and fur clothes. Russian vodka and caviar are among the food that attracts foreigners.

The warm subtropical Black Sea coast of Russia is the site for a number of popular sea resorts, like Sochi, the follow-up host of the 2014 Winter Olympics. The mountains of the Northern Caucasus contain popular ski resorts, including Dombay. The most famous natural destination in Russia is Lake Baikal, *the Blue Eye of Siberia*. This unique lake, oldest and deepest in the world,[] has crystal-clean waters and is surrounded by taiga-covered mountains. Other popular natural destinations include Kamchatka with its volcanoes and geysers, Karelia with its lakes and granite rocks, the snowy Altai Mountains, and the wild steppes of Tyva.

See also

- Culture of Russia
- Geology of Russia
- Index of Soviet Union-related articles
- International rankings of Russia
- Outline of Russia
- Timeline of Russian history

References

[1] http://www.gks.ru/bgd/free/B13_00/IssWWW.exe/Stg/dk04/8-0.htm

[2] http://data.worldbank.org/indicator/NY.GDP.MKTP.PP.CD?order=wbapi_data_value_2012+wbapi_data_value+wbapi_data_value-last&sort=desc

[3] http://data.worldbank.org/indicator/NY.GDP.PCAP.PP.CD?order=wbapi_data_value_2012+wbapi_data_value+wbapi_data_value-last&sort=desc

[4] http://data.worldbank.org/indicator/NY.GDP.MKTP.CD

[5] hhttp://data.worldbank.org/indicator/NY.GDP.PCAP.CD/countries/1W?display=default

[7] "The names Russian Federation and Russia shall be equal".

[8] **(Russian)** Official estimate (http://www.gks.ru/bgd/free/B12_00/IssWWW.exe/Stg/dk10/8-0.htm) as of 1 September 2012.

[9] *The Mongol empire: its rise and legacy*, By Michael Prawdin, Gérard Chaliand, (2005) page 512-550

[11] Peter Turchin, Thomas D. Hall and Jonathan M. Adams, " East-West Orientation of Historical Empires (http://jwsr.ucr.edu/archive/vol12/number2/pdf/jwsr-v12n2-tah.pdf)", *Journal of World-Systems Research* Vol. 12 (no. 2), pp. 219–229 (2006).

[14] Rozhnov, Konstantin, Who won World War II? (http://news.bbc.co.uk/2/hi/europe/4508901.stm). BBC.

[16] International Energy Agency – Oil Market Report 18 January 2012 (http://omrpublic.iea.org/omrarchive/18jan12sup.pdf). Retrieved 20 February 2012.

[17] Country Comparison :: Natural gas – production (https://www.cia.gov/library/publications/the-world-factbook/rankorder/2180rank.html). CIA World Factbook. Retrieved 17 February 2012.

[38] The history of banya and sauna (http://web.archive.org/web/20121130084214/http://sauna-banya.ru/ist.html) **(Russian)**

[39] " *Black Death* (http://books.google.com/books?id=yw3HmjRvVQMC&pg=PA62)". Joseph Patrick Byrne (2004). p. 62. ISBN 0-313-32492-1

[41] Frank D. McConnell. Storytelling and Mythmaking: Images from Film and Literature. (http://books.google.com/books?id=rqhZAAAAMAAJ) Oxford University Press, 1979. ISBN 0-19-502572-5; p. 78: "But Ivan IV, Ivan the Terrible, or as the Russian has it, *Ivan Groznyi*, "Ivan the Magnificent" or "Ivan the Awesome," is precisely a man who has become a legend"

[52] Geoffrey A. Hosking (2001). " *Russia and the Russians: a history* (http://books.google.com/books?id=oh-5AAmboMUC&pg=PA9)". Harvard University Press. p. 9. ISBN 0-674-00473-6

[53] N. M. Dronin, E. G. Bellinger (2005). " *Climate dependence and food problems in Russia, 1900–1990: the interaction of climate and agricultural policy and their effect on food problems* (http://books.google.com/books?id=9a5j_JL6cqIC&pg=PA38)". Central European University Press. p. 38. ISBN 963-7326-10-3

[54] Transactions of the American Philosophical Society (http://books.google.com/books?id=uUsLAAAAIAAJ&pg=PA3). James E. Hassell (1991), p. 3. ISBN 0-87169-817-X

[55] Famine in Russia: the hidden horrors of 1921 (http://web.archive.org/web/20121219101057/http://www.icrc.org/eng/resources/documents/misc/5rfhjy.htm), International Committee of the Red Cross

[56] Abbott Gleason (2009). *A Companion to Russian History* (http://books.google.com/books?id=JyN0hlKcfTcC&pg=PA373). Wiley-Blackwell. p. 373. ISBN 1-4051-3560-3

[57] Getty, Rittersporn, Zemskov. Victims of the Soviet Penal System in the Pre-War Years: A First Approach on the Basis of Archival Evidence. The American Historical Review, Vol. 98, No. 4 (Oct. 1993), pp. 1017–49.

[58] R.W. Davies, S.G. Wheatcroft (2004). *The Years of Hunger: Soviet Agriculture, 1931–33*, pp. 401.

[61] The Legacy of the Siege of Leningrad, 1941–1995 (http://www.cambridge.org/gb/knowledge/isbn/item1173696/?site_locale=en_GB). Cambridge University Press.

[63] Geoffrey A. Hosking (2006). " *Rulers and victims: the Russians in the Soviet Union* (http://books.google.com/books?id=CDMVMqDvp4QC&pg=PA242)". Harvard University Press. p. 242. ISBN 0-674-02178-9

[65] Foreign trade (http://rs6.loc.gov/cgi-bin/query/r?frd/cstdy:@field(DOCID+su0391)) from *A Country Study: Soviet Union (Former)*. Library of Congress Country Studies project.

[80] Venezuela's President Hugo Chavez recognizes independence of breakaway Georgia republics (http://articles.latimes.com/2009/sep/11/world/fg-russia-chavez11) by Megan K. Stack. 9 September 2009

[81] Netanyahu declares Russia as superpower (http://freevideo.rt.com/video/1759) Russia Today News 15 February 2010

[82] Superpower Reborn (http://georgiandaily.com/index.php?option=com_content&task=view&id=6527) by Ronald Steel. New York Times, 24 August 2008
[83] Russia is a Superpower CNN, US Senators telling the truth (http://video.google.com/videoplay?docid=-9079543725663390621) CNN News 30 August 2008
[84] Steven Rosefielde, Russia in the 21st Century The Prodigal Superpower (http://books.google.com/books?id=eC6HdSYZhRgC), Cambridge University Press, 2005 ISBN 0521836786
[85] Is Russia a Superpower? Cold War II? (http://atlanticreview.org/archives/1160-Is-Russia-a-Superpower-Cold-War-II.html) Atlantic Review, 25 August 2008.
[86] What's Looming in Ukraine Is more Threatening than Georgia (http://www.spiegel.de/international/world/0,1518,584631,00.html) Der Spiegel, 16 October 2008. Quote: "Nikonov: Russia is not a superpower and won't be one for the foreseeable future. But Russia is a great power. It was one, it is one and it will continue to be one."
[88] Human Rights Watch on Russia and Chechnya HTW.org (http://www.hrw.org/en/video/2008/04/06/russia-trial)
[91] The Shanghai Cooperation Organisation (http://www.globalsecurity.org/military/world/int/sco.htm) at Globalsecurity.org 27 April 2005
[92] (Russian)
[95] Russia pilots proud of flights to foreign shores (http://www.komonews.com/news/national/28390779.html) by David Nowak. The Associated Press, 15 September 2008
[103] Direct elections of heads of republics (and other federal subjects) are mandated by art. 18 of the Federal Law of 06.10.1999 № 184-FZ (http://constitution.garant.ru/act/federative/12117177/chapter/3/#300) as amended by Federal Law of 2.05.2012 № 40-FZ (http://base.garant.ru/70169404/#131)
[104] Designation of republican heads as presidents is forbidden by Federal Law of 28.12.2010 № 406-FZ (http://graph.document.kremlin.ru/page.aspx?1;1538006), but transitional period lasts until 1 January 2015
[105] *Russian Classification of Economic Regions* (OK 024–95) of 1 January 1997 as amended by the Amendments #1/1998 through #5/2001. (Section I. Federal Districts)
[122] Debt – external (https://www.cia.gov/library/publications/the-world-factbook/rankorder/2079rank.html), CIA World Factbook. Retrieved 22 May 2010.
[127] GRP by federal subjects of Russia, 1998–2007 (http://www.gks.ru/bgd/free/b01_19/IssWWW.exe/Stg/d000/vrp98-07.htm) (Russian)
[131] Land Use (https://www.cia.gov/library/publications/the-world-factbook/fields/2097.html) CIA World Factbook
[132] Data (http://www.gks.ru/bgd/regl/b09_02/IssWWW.exe/Stg/d010/1-03.doc) by Rosstat (Russian)
[133] Russia takes the third place in the world by grain exports (http://web.archive.org/web/20120108052630/http://www.rosbankjournal.ru/news/11588), rosbankjournal.ru (Russian)
[134] Data (http://www.gks.ru/bgd/regl/b09_02/IssWWW.exe/Stg/d010/1-04.doc) by Rosstat (Russian)
[135] Agricultural land by type of owners (http://www.gks.ru/bgd/regl/b09_38/IssWWW.exe/Stg/d01/04-01.htm) Rosstat, 2009 (Russian)
[136] Main agricultural products by type of owners (http://www.gks.ru/bgd/regl/b09_38/IssWWW.exe/Stg/d01/02-28.htm) Rosstat, 2009 (Russian)
[138] Exports and imports of fish and sea products (http://www.gks.ru/bgd/regl/b09_13/IssWWW.exe/Stg/html4/15-07.htm) Rosstat, 2009 (Russian)
[139] FAO. 2010. Global Forest Resources Assesment 2010. Main Report. FAO Forestry Working Paper 163, Rome, Italy (http://www.fao.org/docrep/014/i1757r/i1757r.pdf) (Russian)
[142] FAO. 2012. Russian Federation Forest Sector Outlook Study to 2030. Rome, Italy (http://www.fao.org/docrep/016/i3020r/i3020r00.pdf) (Russian)
[143] Russia, China in Deal On Refinery, Not Gas (http://online.wsj.com/article/SB10001424052748704129204575505373365685564.html) by Jacob Gronholt-Pedersen. Wall Street Journal, 22 September 2010
[144] Did A New Pipeline Just Make Russia The Most Important Energy Superpower By Far (http://www.businessinsider.com/why-russia-is-about-to-change-the-world-2010-1) by Graham Winfrey. Business Insider, 6 January 2010
[145] Country Comparison :: Natural gas – proved reserves (https://www.cia.gov/library/publications/the-world-factbook/rankorder/2179rank.html). CIA World Factbook. Retrieved 5 October 2010.
[146] Country Comparison :: Oil – proved reserves (https://www.cia.gov/library/publications/the-world-factbook/rankorder/2178rank.html). CIA World Factbook. Retrieved 5 October 2010.
[148] Country Comparison :: Natural gas – exports (https://www.cia.gov/library/publications/the-world-factbook/rankorder/2183rank.html). CIA World Factbook. Retrieved 5 October 2010.
[150] Country Comparison :: Electricity – production (https://www.cia.gov/library/publications/the-world-factbook/rankorder/2038rank.html). CIA World Factbook. Retrieved 5 October 2010.
[152] Nuclear Power Plant Information (http://www.iaea.org/programmes/a2/index.html), International Atomic Energy Agency. Retrieved 12 June 2006.
[153] Russia builds nuclear power stations all over the world (http://amur.kp.ru/daily/24458/620649/) at amur.kp.ru (Russian)

[155] CIS railway timetable (http://www.poezda.net/en/web), route No. 002, Moscow-Pyongyang, August 2009. Note: several different routes have the same number.
[156] CIS railway timetable (http://www.poezda.net/en/web), route No. 350, Kiev-Vladivostok, August 2009.
[157] Rosstat statistics on length of roads (http://www.gks.ru/bgd/regl/b08_11/IssWWW.exe/Stg/d02/18-09.htm) Retrieved 10 June 2009
[159] http://englishrussia.com/index.php/2008/03/31/russian-atomic-icebreakers/
[164] Panzerkampfwagen T-34(r) (http://www.achtungpanzer.com/panzerkampfwagen-t-34r-soviet-t-34-in-german-service.htm) by George Parada (n.d.) *Achtung Panzer!* website. Retrieved 17 November 2008
[165] Halberstadt, Hans Inside the Great Tanks The Crowood Press Ltd. Wiltshire, England 1997 94–96 ISBN 1-86126-270-1: "The T-54/T-55 series is the hands down, all time most popular tank in history."
[167] Medvedev outlines priorities for Russian economy's modernization (http://en.rian.ru/russia/20091011/156428675.html) RIA Novosti
[170] ОБ ИТОГАХ ВСЕРОССИЙСКОЙ ПЕРЕПИСИ НАСЕЛЕНИЯ 2010 ГОДА (*Results of the Russian 2010 census*) (http://www.perepis-2010.ru/results_of_the_census/result-december-2011.ppt) – http://www.perepis-2010.ru/
[171] Ethnic groups in Russia (http://demoscope.ru/weekly/ssp/rus_nac_02.php), 2002 census, Demoscope Weekly. Retrieved 5 February 2009.
[174] Modern demographics of Russia (http://www.gks.ru/free_doc/2010/demo/dem-sit-09.doc) by Rosstat. Retrieved on 5 October 2010
[176] Putin tries to lure millions of Russian expats home (http://www.timesonline.co.uk/tol/news/world/europe/article728982.ece) Times Online. 9 February 2006.
[177] Moya Flynn. Migrant resettlement in the Russian federation: reconstructing 'homes' and 'homelands' (http://books.google.com/books?id=YLeAxHLmgR8C&pg=PA15), Anthem Press (2004). p. 15. ISBN 1-84331-117-8
[182] Russian birth rates 1950–2008 (http://demoscope.ru/weekly/ias/ias05.php?tim=0&cou=26&terr=1&ind=26&Submit=OK) Demoscope Weekly. Retrieved October 2010.
[192] (registration required)
[200] Russian life expectancy figures (http://www.gks.ru/free_doc/new_site/population/demo/demo26.xls) Rosstat. Retrieved 21 August 2010
[201] David Johnson, ed., *Politics, Modernisation and Educational Reform in Russia: From Past to Present* (2010)
[207] The first stone tented roof church and the origins of the tented roof architecture (http://web.archive.org/web/20130116214934/http://www.rusarch.ru/zagraevsky19.htm) by Sergey Zagraevsky at RusArch.ru **(Russian)**
[208] The shapes of domes of ancient Russian churches (http://web.archive.org/web/20130116215900/http://www.rusarch.ru/zagraevsky1.htm) by Sergey Zagraevsky at the site of RusArch.ru **(Russian)**
[209] Russian: Постановление ЦК КПСС и СМ СССР "Об устранении излишеств в проектировании и строительстве", 04.11.1955 (Khrushchev's decree *On liquidation of excesses...*) **(Russian)**
[210] Over 20,000 churches rebuilt in Russia in 20 years – Patriarch Kirill (http://www.en.rian.ru/art_living/20101203/161616486.html) RIA Novosti
[211] Russian Academy of Arts (http://en.rah.ru/content/en/section-academy-en.html) official site.
[218] History of Rock Music in Russia (http://www.russia-ic.com/culture_art/music/380/#r1) at Russia-InfoCentre
[223] Birgit Beumers. *A History of Russian Cinema*. Berg Publishers (2009). ISBN 978-1-84520-215-6. p. 143.

External links

- Russia (http://www.dmoz.org/Regional/Europe/Russia/) at the Open Directory Project
- Wikimedia Atlas of Russia
- Russia (https://www.cia.gov/library/publications/the-world-factbook/geos/rs.html) entry at *The World Factbook*
- Russia (http://ucblibraries.colorado.edu/govpubs/for/russia.htm) at *UCB Libraries GovPubs*
- Russia (http://www.bbc.co.uk/news/world-europe-17839672) from the BBC News
- Russia (http://www.britannica.com/EBchecked/topic/513251/Russia) at *Encyclopædia Britannica*

Government

- Official Russian governmental portal (http://www.gov.ru/index_en.html)
- Chief of State and Cabinet Members (https://www.cia.gov/library/publications/world-leaders-1/world-leaders-r/russia.html)
- Russian News Agency "Ria Novosti" (http://en.rian.ru/)
- Russian radio "Voice of Russia" (http://english.ruvr.ru/2010/09/24/22168374.html)

Other

- Post-Soviet Problems (http://repository.library.georgetown.edu/handle/10822/552653) from the Dean Peter Krogh Foreign Affairs Digital Archives (http://repository.library.georgetown.edu/handle/10822/552494/browse?type=title)
- Russia Beyond the Headlines (http://rbth.ru/) International news project about Russia
- Way to Russia. An Introduction to Russia and Russian People (http://www.waytorussia.net/WhatIsRussia/Intro.html)
- Russia cities and regions guide (http://russiatrek.org/)
- Official Russia Travel Guide (http://www.russia-travel.com/)
- Russian Consulate (http://russianconsulate.com/)
- Russia Beyond the Headlines (http://rbth.ru/) International news project about Russia
- Moscow Russia Insider's Guide (http://www.moscow-russia-insiders-guide.com/) Moscow and Russia through Muscovite's eyes.

Evenk_Autonomous_Okrug

Evenk Autonomous Okrug (Russian: Эвенки́йский автоно́мный о́круг, *Evenkiysky avtonomny okrug*; Evenki: Эведы Автомоды Округ, *Evedy avtomody okrug*), or **Evenkia**, was a federal subject of Russia (an autonomous okrug of Krasnoyarsk Krai). It had been created in 1930. Its administrative center was the urban-type settlement of Tura. As of 2006, at 767,600 km², it was Russia's seventh largest federal subject, and the country's least populous: 17,697 (2002 Census).[]

Flag of Evenk Autonomous Okrug

In 1999, the governor of Krasnoyarsk, General Alexander Lebed, demanded the okrug recognize the central district government of Krasnoyarsk had authority over it, which the okrug refused to do, causing a power struggle between the central district and the okrug's government. [1]

Following a referendum on the issue held on April 17, 2005, Evenk and Taymyr Autonomous Okrugs were merged into Krasnoyarsk Krai effective January 1, 2007 (some Evenks contested the results, however). Administratively, they are now considered to be districts with special status within Krasnoyarsk Krai; municipally, they have a status of municipal districts (see Evenkiysky District).

Boris Zolotaryov was the last governor of the autonomous okrug.

Demographics

Population

(2002): 17,697.

Vital statistics

Source: Russian Federal State Statistics Service [6]

	Average population (x 1000)	Live births	Deaths	Natural change	Crude birth rate (per 1000)	Crude death rate (per 1000)	Natural change (per 1000)
1970	13	314	144	170	24.2	11.1	13.1
1975	15	254	159	95	16.9	10.6	6.3
1980	17	373	167	206	21.9	9.8	12.1
1985	22	521	219	302	23.7	10.0	13.7
1990	24	514	189	325	21.3	7.8	13.5
1991	24	427	221	206	17.7	9.1	8.5
1992	24	414	249	165	17.5	10.5	7.0
1993	23	297	270	27	13.1	11.9	1.2
1994	21	294	257	37	13.9	12.1	1.7
1995	20	299	214	85	14.8	10.6	4.2
1996	20	269	223	46	13.5	11.2	2.3
1997	20	261	202	59	13.3	10.3	3.0
1998	19	244	220	24	12.7	11.4	1.2
1999	19	251	203	48	13.4	10.8	2.6
2000	18	242	214	28	13.3	11.7	1.5
2001	18	274	234	40	15.3	13.1	2.2
2002	18	263	237	26	14.9	13.4	1.5
2003	18	274	215	59	15.6	12.3	3.4
2004	17	267	218	49	15.4	12.6	2.8
2005	17	283	259	24	16.5	15.1	1.4
2006	17	282	236	46	16.6	13.9	2.7

Ethnic groups

Of the 17,697 residents (as of the 2002 Census) 2 (0.01%) chose not to specify their ethnic background. Of the rest, residents identified themselves as belonging to 67 ethnic groups, including ethnic Russians (62%), Evenks (21.5%), Yakuts (5.6%), Ukrainians (3.1%), Kets (1.2%), 162 Tatars (0.9%), 152 Khakas (0.9%) and 127 Volga Germans (0.7%).

Ethnic group	1939 census		1959 census		1970 census		1979 census		1989 census		2002 census	
	Number	%	Number	%	Number	%	Number	%	Number	%	Number	%
Evenks	3,721	39.3%	3,474	33.7%	3,207	25.3%	3,239	20.3%	3,480	14.0%	3,802	21.5%
Yakuts	713	7.5%	51	0.5%	781	6.2%	822	5.1%	937	3.8%	991	5.6%
Kets			14	0.1%	142	1.1%	154	1.0%	150	0.6%	211	1.2%
Russians	4,675	49.4%	5,975	57.9%	7,732	61.1%	10,400	65.1%	16,718	67.5%	10,958	61.9%
Ukrainians	117	1.2%	196	1.9%	254	2.0%	472	3.0%	1,303	5.3%	550	3.1%
Others	234	2.5%	610	5.9%	542	4.3%	881	5.5%	2,181	8.8%	1,185	6.7%

References

[1] National Geographic. *The Book of Peoples*. Page 139

See also

- Music of Evenkia

Federal_subjects_of_Russia

The **federal subjects of Russia**, also referred to as the **subjects of the Russian Federation** (Russian: субъекты Российской Федерации *subyekty Rossiskoy Federatsii*) or simply as the **subjects of the federation** (Russian: субъекты федерации *subyekty federatsii*) are the constituent entities of Russia, its top-level political divisions according to the Constitution of Russia.[1] Since March 1, 2008, the Russian Federation consists of eighty-three federal subjects.[]

In 1992, the Russian regions signed the Federation Treaty (Russian: Федеративный договор *Federativny Dogovor*),[2] establishing and regulating the current inner composition of Russia, based on division of authorities and powers among Russian government bodies and government bodies of constituent entities. The Federation Treaty was included in the text of the 1978 Constitution of the RSFSR. In 1993, when the current Constitution of Russia was adopted, there were 89 federal subjects listed. By 2008, the number of federal subjects had been decreased to 83 because of several mergers. 2 Russian cities of federal importance have a status of both city and separated federal subject.

According to the Constitution, subjects shall have equal rights as constituent entities of the Russian Federation. They shall have their own constitution/charter and legislation (adopted by regional parliaments). All federal subjects of the Russian Federation shall be equal with one another in relations with federal State government bodies.[3] [4] The federal subjects have equal representation—two delegates each—in the Federation Council, the upper house of the Federal Assembly. They do, however, differ in the degree of autonomy they enjoy (asymmetric federalism).

Federal subjects should not be confused with the eight *Federal districts* which are not subdivisions of Russia, are much larger and each encompass many federal subjects. Federal districts were created by Executive Order of the President of Russia specially for presidential envoys.

Terminology

The official government translation of the Constitution of Russia in Article 5 states:[5] "1. The Russian Federation shall consist of republics, krays, oblasts, cities of federal significance, an autonomous oblast and autonomous okrugs, which shall have equal rights as constituent entities of the Russian Federation."

How to translate the Russian term was discussed during the 49th annual American Translators Association conference in Orlando, in which Tom Fennel, a freelance translator, argued that the term "constituent entity of the Russian Federation" should be preferred to "subject".[6] This recommendation is also shared by Tamara Nekrasova, Head of Translation Department, Goltsblat BLP, who in her "Traps & Mishaps in Legal Translation" presentation in Paris stated that "*constituent entity of the Russian Federation* is more appropriate than *subject of the Russian Federation* (subject would be OK for a monarchy)".[7]

Types

Each federal subject belongs to one of the following types:

21 republics (республика, *respublika*) — nominally autonomous, each has its own constitution and legislature; is represented by the federal government in international affairs; is meant to be home to a specific ethnic minority.

46 oblasts (provinces; область, *oblast*) — most common type of federal subjects with federally appointed governor and locally elected legislature. Commonly named after their administrative centers.

9 krais (territories; край, *kraĭ*)—essentially the same as oblasts. The title "territory" is historic, originally given because they were once considered frontier regions.

1 autonomous oblast (autonomous province; автономная область, *avtonomnaya oblast*)—the only autonomous oblast is the Jewish Autonomous Oblast

4 autonomous okrugs (autonomous districts; автономный округ, *avtonomnyĭ okrug*) — with substantial or predominant ethnic minority

2 federal cities (город федерального значения, *gorod federal'nogo znacheniya*) — major cities that function as separate regions.

List of federal subjects

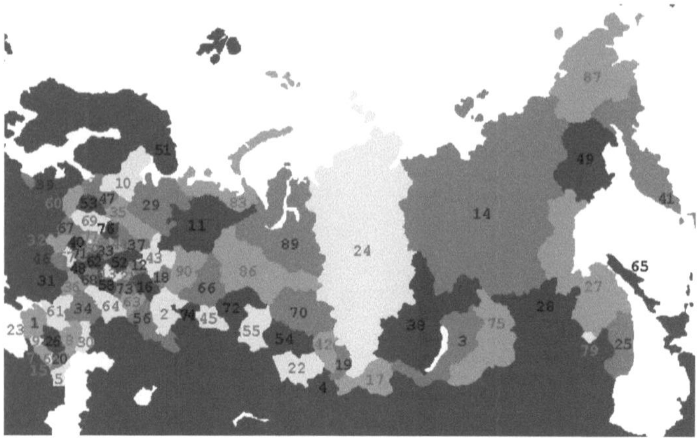

The subjects have both numerical codes and two- or three-letter ISO 3166-2:RU codes. The numerical codes span from 01 to 92, although nine of them (41, 59, 75, 80, 81, 82, 84, 85, and 88) are no longer in use after mergers.

Code	Name	Capital/administrative centre (Largest city given if not capital)	Flag	Coat of arms	Federal district	Economic region	Area (km²)[8]	Population[9]	Establishment Year
01	Adygea, Republic of	Maykop		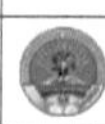	Southern	North Caucasus	7,600	447,109	1922
02	Bashkortostan, Republic of	Ufa			Volga	Ural	143,600	4,104,336	1919
03	Buryatia, Republic of	Ulan-Ude		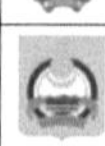	Siberian	East Siberian	351,300	981,238	1923
04	Altai Republic	Gorno-Altaysk		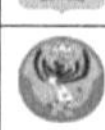	Siberian	West Siberian	92,600	202,947	1922
05	Dagestan, Republic of	Makhachkala	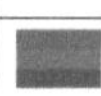		North Caucasian	North Caucasus	50,300	2,576,531	1921
06	Ingushetia, Republic of	Magas (Largest city: Nazran)	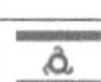		North Caucasian	North Caucasus	4,000	467,294	1992
07	Kabardino-Balkar Republic	Nalchik	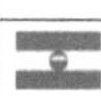		North Caucasian	North Caucasus	12,500	901,494	1936
08	Kalmykia, Republic of	Elista			Southern	Volga	76,100	292,410	1957

09	Karachay-Cherkess Republic	Cherkessk			North Caucasian	North Caucasus	14,100	439,470	1957
10	Karelia, Republic of	Petrozavodsk			Northwestern	Northern	172,400	716,281	1956
11	Komi Republic	Syktyvkar			Northwestern	Northern	415,900	1,018,674	1921
12	Mari El Republic	Yoshkar-Ola			Volga	Volga-Vyatka	23,200	727,979	1920
13	Mordovia, Republic of	Saransk			Volga	Volga-Vyatka	26,200	888,766	1930
14	Sakha (Yakutia) Republic	Yakutsk			Far Eastern	Far Eastern	3,103,200	949,280	1922
15	North Ossetia-Alania, Republic of	Vladikavkaz			North Caucasian	North Caucasus	8,000	710,275	1924
16	Tatarstan, Republic of	Kazan			Volga	Volga	68,000	3,779,265	1920
17	Tuva Republic	Kyzyl			Siberian	East Siberian	170,500	305,510	1944
18	Udmurt Republic	Izhevsk			Volga	Ural	42,100	1,570,316	1920
19	Khakassia, Republic of	Abakan			Siberian	East Siberian	61,900	546,072	1930
20	Chechen Republic	Grozny			North Caucasian	North Caucasus	15,300	1,103,686	1991
21	Chuvash Republic	Cheboksary			Volga	Volga-Vyatka	18,300	1,313,754	1920
22	Altai Krai	Barnaul			Siberian	West Siberian	169,100	2,607,426	1937
92	Zabaykalsky Krai	Chita			Siberian	East Siberian	431,500	1,155,346	2008
91	Kamchatka Krai	Petropavlovsk-Kamchatsky			Far Eastern	Far Eastern	472,300	358,801	2007
23	Krasnodar Krai	Krasnodar			Southern	North Caucasus	76,000	5,125,221	1937
24	Krasnoyarsk Krai	Krasnoyarsk			Siberian	East Siberian	2,339,700	2,966,042	1934
90	Perm Krai	Perm			Volga	Ural	160,600	2,819,421	2005
25	Primorsky Krai	Vladivostok			Far Eastern	Far Eastern	165,900	2,071,210	1938
26	Stavropol Krai	Stavropol			North Caucasian	North Caucasus	66,500	2,735,139	1934
27	Khabarovsk Krai	Khabarovsk			Far Eastern	Far Eastern	788,600	1,436,570	1938

28	Amur Oblast	Blagoveshchensk			Far Eastern	Far Eastern	363,700	902,844	1932
29	Arkhangelsk Oblast	Arkhangelsk			Northwestern	Northern	587,400	1,336,539	1937
30	Astrakhan Oblast	Astrakhan			Southern	Volga	44,100	1,005,276	1943
31	Belgorod Oblast	Belgorod			Central	Central Black Earth	27,100	1,511,620	1954
32	Bryansk Oblast	Bryansk			Central	Central	34,900	1,378,941	1944
33	Vladimir Oblast	Vladimir			Central	Central	29,000	1,523,990	1944
34	Volgograd Oblast	Volgograd			Southern	Volga	113,900	2,699,223	1937
35	Vologda Oblast	Vologda (Largest city: Cherepovets)			Northwestern	Northern	145,700	1,269,568	1937
36	Voronezh Oblast	Voronezh			Central	Central Black Earth	52,400	2,378,803	1934
37	Ivanovo Oblast	Ivanovo			Central	Central	21,800	1,148,329	1936
38	Irkutsk Oblast	Irkutsk			Siberian	East Siberian	767,900	2,581,705	1937
39	Kaliningrad Oblast	Kaliningrad			Northwestern	Kaliningrad	15,100	955,281	1946
40	Kaluga Oblast	Kaluga			Central	Central	29,900	1,041,641	1944
42	Kemerovo Oblast	Kemerovo (Largest city: Novokuznetsk)			Siberian	West Siberian	95,500	2,899,142	1943
43	Kirov Oblast	Kirov			Volga	Volga-Vyatka	120,800	1,503,529	1934
44	Kostroma Oblast	Kostroma			Central	Central	60,100	736,641	1944
45	Kurgan Oblast	Kurgan			Ural	Ural	71,000	1,019,532	1943
46	Kursk Oblast	Kursk			Central	Central Black Earth	29,800	1,235,091	1934
47	Leningrad Oblast	Largest city: Gatchina[a]			Northwestern	Northwestern	84,500	1,669,205	1927
48	Lipetsk Oblast	Lipetsk			Central	Central Black Earth	24,100	1,213,499	1954
49	Magadan Oblast	Magadan			Far Eastern	Far Eastern	461,400	182,726	1953
50	Moscow Oblast	Largest city: Balashikha[b]			Central	Central	45,900	6,618,538	1929

51	Murmansk Oblast	Murmansk			Northwestern	Northern	144,900	892,534	1938
52	Nizhny Novgorod Oblast	Nizhny Novgorod			Volga	Volga-Vyatka	76,900	3,524,028	1936
53	Novgorod Oblast	Veliky Novgorod			Northwestern	Northwestern	55,300	694,355	
54	Novosibirsk Oblast	Novosibirsk			Siberian	West Siberian	178,200	2,692,251	1937
55	Omsk Oblast	Omsk			Siberian	West Siberian	139,700	2,079,220	
56	Orenburg Oblast	Orenburg			Volga	Ural	124,000	2,179,551	
57	Oryol Oblast	Oryol			Central	Central	24,700	860,262	
58	Penza Oblast	Penza			Volga	Volga	43,200	1,452,941	
60	Pskov Oblast	Pskov			Northwestern	Northwestern	55,300	760,810	
61	Rostov Oblast	Rostov-on-Don			Southern	North Caucasus	100,800	4,404,013	
62	Ryazan Oblast	Ryazan			Central	Central	39,600	1,227,910	
63	Samara Oblast	Samara			Volga	Volga	53,600	3,239,737	
64	Saratov Oblast	Saratov			Volga	Volga	100,200	2,668,310	
65	Sakhalin Oblast	Yuzhno-Sakhalinsk			Far Eastern	Far Eastern	87,100	546,695	
66	Sverdlovsk Oblast	Yekaterinburg			Ural	Ural	194,800	4,486,214	
67	Smolensk Oblast	Smolensk			Central	Central	49,800	1,049,574	
68	Tambov Oblast	Tambov			Central	Central Black Earth	34,300	1,178,443	
69	Tver Oblast	Tver			Central	Central	84,100	1,471,459	
70	Tomsk Oblast	Tomsk			Siberian	West Siberian	316,900	1,046,039	
71	Tula Oblast	Tula			Central	Central	25,700	1,675,758	
72	Tyumen Oblast	Tyumen			Ural	West Siberian	1,435,200	3,264,841	
73	Ulyanovsk Oblast	Ulyanovsk			Volga	Volga	37,300	1,382,811	

74	Chelyabinsk Oblast	Chelyabinsk			Ural	Ural	87,900	3,603,339	
76	Yaroslavl Oblast	Yaroslavl			Central	Central	36,400	1,367,398	
77	Moscow	—			Central	Central	1,100	10,382,754	
78	Saint Petersburg	—			Northwestern	Northwestern	1,439	4,662,547	
79	Jewish Autonomous Oblast	Birobidzhan			Far Eastern	Far Eastern	36,000	190,915	
83	Nenets Autonomous Okrug	Naryan-Mar			Northwestern	Northern	176,700	41,546	
86	Khanty–Mansi Autonomous Okrug – Yugra	Khanty-Mansiysk (Largest city: Surgut)			Ural	West Siberian	523,100	1,432,817	
87	Chukotka Autonomous Okrug	Anadyr			Far Eastern	Far Eastern	737,700	53,824	
89	Yamalo-Nenets Autonomous Okrug	Salekhard (Largest city: Novy Urengoy)			Ural	West Siberian	750,300	507,006	

Mergers

Federal subjects of Russia

Starting in 2005, some of the federal subjects were merged into larger territories. The merging process was finished on March 1, 2008. No new mergers have been planned since March 2008.

Original territories	Original codes	New code	Date of referendum	Date of merger	Merger
1, 1a	59 (1), 81 (1a)	90	December 7, 2003	December 1, 2005	Perm Oblast (1) + Komi-Permyak Autonomous Okrug (1a) → Perm Krai
2, 2a, 2b	24 (2), 88 (2a), 84 (2b)	24	April 17, 2005	January 1, 2007	Krasnoyarsk Krai (2) + Evenk Autonomous Okrug (2a) + Taymyr Autonomous Okrug (2b) → Krasnoyarsk Krai
3, 3a	41 (3), 82 (3a)	91	October 23, 2005	July 1, 2007	Kamchatka Oblast (3) + Koryak Autonomous Okrug (3a) → Kamchatka Krai
4, 4a	38 (4), 85 (4a)	38	April 16, 2006	January 1, 2008	Irkutsk Oblast (4) + Ust-Orda Buryat Autonomous Okrug (4a) → Irkutsk Oblast
5, 5a	75 (5), 80 (5a)	92	March 11, 2007	March 1, 2008	Chita Oblast (5) + Agin-Buryat Autonomous Okrug (5a) → Zabaykalsky Krai

Further proposals for mergers

The following merger proposals have been made in recent years; most have since become inactive.

Proposals	Federal subjects	Map
6, 6a	Arkhangelsk Oblast + Nenets Autonomous Okrug = Pomorsky Krai	
7, 7a, 7b	Khabarovsk Krai + Jewish Autonomous Oblast + Amur Oblast = Amur Krai	
8, 8a	Magadan Oblast + Chukotka Autonomous Okrug = Magadan Krai	
9, 9a, 9b	Irkutsk Oblast + Buryat Republic + Zabaykalsky Krai = Baykalsky Krai	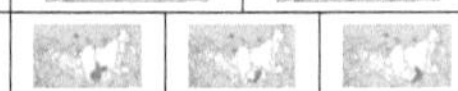
10, 10a, 10b	Tyumen Oblast + Yamalo-Nenets Autonomous Okrug + Khanty–Mansi Autonomous Okrug = Tyumen Krai	
11, 11a, 11b	Novosibirsk Oblast + Omsk Oblast + Tomsk Oblast = Novosibirsk Krai	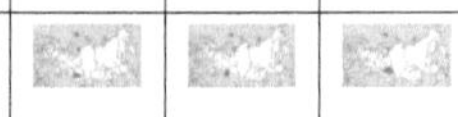
12, 12a, 12b	Kemerovo Oblast + Altai Republic + Altai Krai = Altai Krai	
13	St. Petersburg + Leningrad Oblast = St. Petersburg Oblast	
14	Moscow + Moscow Oblast = Moscow Oblast	
15, 15a	Yaroslavl Oblast + Kostroma Oblast = Yaroslavl Krai	
16, 16a	Novgorod Oblast + Pskov Oblast = Novgorod Krai	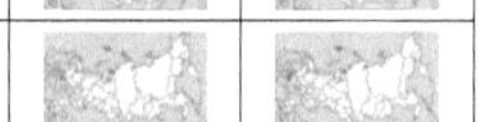
17, 17a	Krasnodar Krai + Republic of Adygea = Krasnodar Krai	
18, 18a	Republic of Ingushetia + Chechen Republic = Checheno-Ingushetia	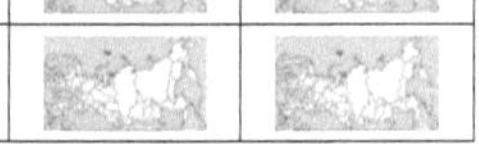

See also

- Subdivisions of Russia
- Federal districts of Russia
- Economic regions of Russia
- History of the administrative division of Russia
- Flags of the federal subjects of Russia
- List of heads of federal subjects of Russia
- List of Russian federal subjects by GRP

References

[2] this treaty consisted of 3 treaties, see also Concluding and Transitional Provisions: (http://eng.constitution.kremlin.ru/#conclusion-1-4) (http://www.constitution.ru/en/10003000-10.htm#2222)
[3] http://eng.constitution.kremlin.ru/#article-5
[4] http://www.constitution.ru/en/10003000-02.htm
[5] http://www.government.ru/eng/gov/base/54.html
[6] http://www.ata-divisions.org/SLD/slavfile/winter-2009.pdf
[7] http://eulita.eu/sites/default/files/Tammy_presentation.pdf

Sources

• 12 декабря 1993 г. « Конституция Российской Федерации (http://www.consultant.ru/online/base/?req=doc;base=LAW;n=83686)», в ред. Федерального конституционного закона №7-ФКЗ от 30 декабря 2008 г. Вступил в силу со дня официального опубликования. Опубликован: "Российская газета", №237, 25 декабря 1993 г. (December 12, 1993 *Constitution of the Russian Federation* (http://translate.google.com/translate?hl=en&sl=ru&tl=en&u=http://www.consultant.ru/online/base/?req=doc;base=LAW;n=83686), as amended by the Federal Constitutional Law #7-FKZ of December 30, 2008. Effective as of the official publication date.).

Administrative_centre

An **administrative centre** is a seat of regional administration or local government, or a county town, or the place where the central administration of a commune is located.

In Russia, the term is applied to the inhabited localities which serve as a seat of government of entities of various levels. The only exception to this rule is the republics, for which the term "capital" is used to refer to the seat of government. The capital of Russia is also an entity to which the term "administrative centre" does not apply. A similar arrangement exists in Ukraine.

In the United Kingdom it is the centre of a local authority which is distinct from a historic county with a county town.

See also

- Local government
- Chef-lieu, administrative centres in Algeria, Belgium, Luxembourg, France, Switzerland, and Tunisia
- County seat, administrative centres in the USA
- County town, administrative centres in Ireland and the UK
- List of regions of Norway
- Administrative divisions of Portugal
- Subdivisions of Russia

References

Types_of_inhabited_localities_in_Russia

The classification system of the types of inhabited localities in Russia, the former Soviet Union, and some other post-Soviet states has certain peculiarities compared to the classification systems in other countries.

Modern classification in Russia

During the Soviet time, each of the republics of the Soviet Union, including the Russian SFSR, had its own legislative documents dealing with classification of inhabited localities.[1] After the dissolution of the Soviet Union, the task of developing and maintaining such classification in Russia was delegated to the federal subjects.[2] While currently there are certain peculiarities to classifications used in many federal subjects, they are all still largely based on the system used in the RSFSR. In all federal subjects, the inhabited localities are classified into two major categories: urban and rural.[3] Further divisions of these categories vary slightly from one federal subject to another,[2] but they all follow common trends described below.

Urban localities

- **Cities** and **towns** (город, *gorod*; pl. города, *goroda*). Cities and towns are classified by their level of jurisdiction (district/federal subject/federal). While the Russian language has no separate words for "town" and "city" ("город" is used for both), in translation, the word "city" is traditionally applied to the urban localities with a population of at least 100,000.
- **Urban-type settlements** (посёлок городского типа, *posyolok gorodskogo tipa*; pl. посёлки городского типа) is a type of smaller urban locality. This type of urban locality was first introduced in the Soviet Union in 1924, with the following subcategories:[4]
  - Urban-type settlement proper—mostly urban population of 3,000–12,000.
    - Work settlement (рабочий посёлок, *rabochy posyolok*)—mostly urban population occupied in industrial manufacture.
    - Suburban (dacha) settlement (дачный посёлок, *dachny posyolok*)—typically, a suburban settlement with summer dachas.
    - Resort settlement (курортный посёлок, *kurortny posyolok*)—mostly urban population occupied in services.

In 1957, the procedures for categorizing urban-type settlements were further refined.[5]

Rural localities

Multiple types of rural localities exist, some common through the whole territory of Russia, some specific to certain federal subjects. The most common types include:

- **Villages** (деревня, *derevnya*; pl. деревни, *derevni*)
- **Selos** (село, *selo*; pl. сёла, *syola*), also translated as village
- **(Rural-type) settlements** (посёлок (сельского типа), *posyolok (selskogo tipa)*; pl. посёлки (сельского типа)). The "rural-type" (сельского типа) designation is added to the settlements the population of which is mostly occupied in agriculture, while *posyolok* (посёлок) proper indicates a mix of population working in agriculture and industry.
- **Stanitsas** (станица, *stanitsa*; pl. станицы, *stanitsy*)—historically, a Cossack rural locality. The name is still currently in use, with the basic meaning of "village".
- **Slobodas** (слобода, *sloboda*; pl. слободы, *slobody*)—historically, a settlement freed from taxes and levies for various reasons. The name is still currently in use with the basic meaning of "village".
- **Khutors** (хутор, *khutor*; pl. хутора, *khutora*)—translated as "hamlet", "farmstead", or "village".

- **Pochinoks** (починок, *pochinok*; pl. починки, *pochinki*)—a newly formed rural locality of one or several families. Pochinoks are established as new settlements and usually grow into larger villages as they develop.
- In some federal subjects, ethnic terminology is used in the Russian language. Such locality types include аул (*aul*), аал (*aal*), and кишлак (*kishlak*).

Historical terms

- **Krepost** (крепость, a fort), a fortified settlement. A Kremlin, Russian citadel, is a major *krepost* usually including a castle and surrounded by posad. **Ostrog**, on the other hand, was a more primitive kind of *krepost* which could be put up quickly within rough walls of debarked pointed timber.
- **Posad** (посад), a medieval suburban settlement.
- **Mestechko** (местечко, from Polish: *miasteczko*; a small town in Western Krai annexed during the Partitions of Poland; typically with Jewish majority.
- **Pogost**

See also

- Subdivisions of Russia
- Town of district significance
- City of federal subject significance
- List of terms for country subdivisions

References

[1] In the Russian SFSR, the issues of the administrative and territorial division, including the system of classification of the inhabited localities, was regulated by the Statute *On Procedure of Resolving the Issues of the Administrative-Territorial Structure of the RSFSR*, approved by the Decree of the Presidium of the Supreme Soviet of the RSFSR on August 17, 1982 (Положение "О порядке решения вопросос административно-территориального устройства РСФСР", утверждённое Указом Президиума Верховного Совета РСФСР от 17 августа 1982 г.)

[2] Articles 71 and 72 of the Constitution of Russia do not name issues of the administrative and territorial structure among the tasks handled on the federal level or jointly with the governments of the federal subjects. As such, all federal subjects pass their own laws establishing the system of the administrative-territorial divisions on their territories.

[3] See, for example, the results of the 2002 population Census (http://www.perepis2002.ru/ct/doc/1_TOM_01_04.xls)

[4] Постановление ВЦИК и СНК РСФСР от 15 сентября 1924 г. "Общее положение о городских и сельских поселениях и посёлках" (Resolution of the All-Union Executive Committee and the Soviet of People's Commissars of September 15, 1924 *General Statute on Urban and Rural Settlements*)

[5] Указ Президиума ВС РСФСР от 12 сентября 1957 г. "О порядке отнесения населённых пунктов к категории городов, рабочих и курортных посёлков" (Decree of the Presidium of the Supreme Soviet of the RSFSR of September 12, 1957 *On Procedures of Categorizing the Inhabited Localities as Cities, Work and Resort Settlements)*

Taymyrsky_Dolgano-Nenetsky_District

Taymyrsky Dolgano-Nenetsky District (Russian: Таймы́рский Долга́но-Не́нецкий райо́н) is an administrative[1] and municipal[2] district (raion), one of the forty-three in Krasnoyarsk Krai, Russia. It is located in the north of the krai above the Arctic Circle on the Taymyr Peninsula and borders with Laptev and Kara Seas in the north, the Sakha Republic in the east, Evenkiysky and Turukhansky Districts in the south, and with Yamalo-Nenets Autonomous Okrug in the west. The area of the district is 879900 square kilometers (sq mi).[3] Its administrative center is the town of Dudinka.[1] Population: 34,432 (2010 Census);[] 39,786 (2002 Census);[] 55,111 (1989 Census).[] The population of Dudinka accounts for 64.4% of the district's total population.[]

History

The district was founded on April 10, 1930.[3] Until January 1, 2007, it existed as Taymyr Autonomous Okrug, a federal subject of Russia.

Government

As of 2013, the Head of the district and the Chairman of the District Duma is Sergey V. Baturin.[3]

Demographics

Vital statistics

Source: Russian Federal State Statistics Service [6]

	Average population (x 1000)	Live births	Deaths	Natural change	Crude birth rate (per 1000)	Crude death rate (per 1000)	Natural change (per 1000)
2007	37	592	335	257	16.2	9.2	7.0
2008	36	586	360	226	16.3	10.0	6.3
2009	35	611	387	224	17.3	11.0	6.4
2010	35	602	351	251	17.4	10.1	7.3

References

Notes

[1] Law #10-4765

[2] Law #2-54

[3] Official website of Krasnoyarsk Krai. Information about Taymyrsky Dolgano-Nenetsky District (http://www.krskstate.ru/msu/terdel/0/doc/58) (Russian)

Sources

- Законодательное собрание Красноярского края. Закон №10-4765 от 10 июня 2010 г. «О перечне административно-территориальных единиц и территориальных единиц Красноярского края», в ред. Закона №13-6663 от 1 декабря 2011 г. «Об изменении статуса рабочих посёлков и внесении изменений в Закон края "О перечне административно-территориальных единиц и территориальных единиц Красноярского края"». Вступил в силу 1 июля 2010 г.. Опубликован: "Ведомости высших органов государственной власти Красноярского края", №33(404), 5 июля 2010 г. (Legislative Assembly of Krasnoyarsk Krai. Law #10-4765 of June 10, 2010 *On the Registry of the Administrative-Territorial Units and the Territorial*

Units of Krasnoyarsk Krai, as amended by the Law #13-6663 of December 1, 2011 *On Changing the Status of the Work Settlements and on Amending the Krai Law "On the Registry of the Administrative-Territorial Units and the Territorial Units of Krasnoyarsk Krai"*. Effective as of July 1, 2010.).

- Законодательное собрание Красноярского края. Закон №2-54 от 9 февраля 2012 г. «Об установлении границ муниципального образования Таймырский Долгано-Ненецкий муниципальный район и находящихся в его границах иных муниципальных образований». Вступил в силу через десять дней со дня официального опубликования. Опубликован: "Ведомости высших органов государственной власти Красноярского края", №8(520)/1 и №8(520)/2, 6 марта 2012 г. (Legislative Assembly of Krasnoyarsk Krai. Law #2-54 of February 9, 2012 *On Establishing the Borders of the Municipal Formation of Taymyrsky Dolgano-Nenetsky Municipal District and of Other Municipal Formations Within Its Borders*. Effective as of the day which is ten days from the day of the official publication.).
- Собрание Таймырского Долгано-Ненецкого муниципального района. Решение №02-0063 от 22 декабря 2005 г. «О принятии Устава Таймырского Долгано-Ненецкого муниципального района», в ред. Решения №03-0039 от 14 апреля 2009 г «О внесении изменений в Устав Таймырского Долгано-Ненецкого муниципального района». Вступил в силу со дня официального опубликования, но не ранее 1 января 2006 года, за исключением положений, для которых пунктами 2 и 3 статьи 90 установлены иные сроки вступления в силу. Опубликован: "Таймыр", №155, 30 декабря 2005 г. (Assembly of Taymyrsky Dolgano-Nenetsky Municipal District. Decision #02-0063 of December 22, 2005 *On the Adoption of the Charter of Taymyrsky Dolgano-Nenetsky Municipal District*, as amended by the Decision #03-0039 of April 14, 2009 *On Amending the Charter of Taymyrsky Dolgano-Nenetsky Municipal District*. Effective as of the day of the official publication, but not earlier than January 1, 2006, with the exception of the clauses for which items 2 and 3 of Article 90 establish different dates of taking effect.).

Severo-Yeniseysky_District

Severo-Yeniseysky District (Russian: Се́веро-Енисе́йский райо́н) is an administrative[1] and municipal[2] district (raion), one of the forty-three in Krasnoyarsk Krai, Russia. It is located in the center of the krai and borders with Evenkiysky District in the northwest, north, and east, Motyginsky District in the southeast, and with Yeniseysky District in the southwest and west. The area of the district is 47242 square kilometers (sq mi).[3] Its administrative center is the urban locality (a work settlement) of Severo-Yeniseysky.[1] Population: 11,119 (2010 Census);[] 11,077 (2002 Census);[] 17,163 (1989 Census).[] The population of the administrative center accounts for 62.5% of the district's total population.[]

History

The district was founded on April 1, 1932.[3]

Government

As of 2013, the Head of the district and the Chairman of the District Council is Ishmurat M. Gaynutdinov.[3]

References

Notes

[1] Law #10-4765
[2] Law #13-3017
[3] Official website of Krasnoyarsk Krai. Information about Severo-Yeniseysky District (http://www.krskstate.ru/msu/terdel/0/doc/31) (Russian)

Sources

- Законодательное собрание Красноярского края. Закон №10-4765 от 10 июня 2010 г. «О перечне административно-территориальных единиц и территориальных единиц Красноярского края», в ред. Закона №13-6663 от 1 декабря 2011 г. «Об изменении статуса рабочих посёлков и внесении изменений в Закон края "О перечне административно-территориальных единиц и территориальных единиц Красноярского края"». Вступил в силу 1 июля 2010 г.. Опубликован: "Ведомости высших органов государственной власти Красноярского края", №33(404), 5 июля 2010 г. (Legislative Assembly of Krasnoyarsk Krai. Law #10-4765 of June 10, 2010 *On the Registry of the Administrative-Territorial Units and the Territorial Units of Krasnoyarsk Krai*, as amended by the Law #13-6663 of December 1, 2011 *On Changing the Status of the Work Settlements and on Amending the Krai Law "On the Registry of the Administrative-Territorial Units and the Territorial Units of Krasnoyarsk Krai"*. Effective as of July 1, 2010.).
- Законодательное собрание Красноярского края. Закон №13-3017 от 18 февраля 2005 г. «Об установлении границ и наделении соответствующим статусом муниципального образования Северо-Енисейский район», в ред. Закона №7-2656 от 18 декабря 2008 г. «О внесении изменений в Закон края «Об установлении границ и наделении соответствующим статусом муниципального образования Северо-Енисейский район»». Вступил в силу через десять дней после официального опубликования. Опубликован: "Ведомости высших органов государственной власти Красноярского края", №9, 4 марта 2005 г. (Legislative Assembly of Krasnoyarsk Krai. Law #13-3017 of February 18, 2005 *On Establishing the Borders and Granting an Appropriate Status to the Municipal Formation of Severo-Yeniseysky District*, as amended by the Law #7-2656 of December 18, 2008 *On Amending the Krai Law "On Establishing the Borders and Granting an Appropriate Status to the Municipal Formation of Severo-Yeniseysky District"*. Effective as of ten days after the official publication.).

Turukhansky_District

Turukhansky District (Russian: Турухáнский райóн) is an administrative[1] and municipal[2] district (raion), one of the forty-three in Krasnoyarsk Krai, Russia. It is located in the west of the krai and borders with Taymyrsky Dolgano-Nenetsky District in the north, Evenkiysky District in the east, Yeniseysky District in the south, and with Tyumen Oblast in the west. The area of the district is 211189 square kilometers (sq mi).[3] Its administrative center is the rural locality (a *selo*) of Turukhansk.[1] Population: 18,708 (2010 Census);[] 12,439 (2002 Census);[] 19,257 (1989 Census).[] The population of Turukhansk accounts for 24.9% of the district's total population.[]

Geography

The following tributaries of the Yenisey flow through the district: the Podkamennaya Tunguska River, the Yeloguy River, the Nizhnyaya Tunguska River, the Turukhan River, and the Kureyka River.

History

The district was founded on June 7, 1928.[3]

Government

As of 2013, the Head of the district and the Chairman of the District Council is Anatoly I. Goloded.[3]

Demographics

The district is home to most of Ket people, a little-known ethnic group whose language is thought by some linguists to be related to the Na-Dene languages of North American Indians. Nowadays, most of people still speaking Ket live in just three localities: Kellog, Surgutikha, and Maduyka, all of which are situated in Turukhansky District.[4]

Notable people

Joseph Stalin lived in exile on the territory of the modern district before the October Revolution.

References

Notes

[1] Law #10-4765
[2] Law #13-2925
[3] Official website of Krasnoyarsk Krai. Information about Turukhansky District (http://www.krskstate.ru/msu/terdel/0/doc/21) (Russian)

Sources

- Законодательное собрание Красноярского края. Закон №10-4765 от 10 июня 2010 г. «О перечне административно-территориальных единиц и территориальных единиц Красноярского края», в ред. Закона №13-6663 от 1 декабря 2011 г. «Об изменении статуса рабочих посёлков и внесении изменений в Закон края "О перечне административно-территориальных единиц и территориальных единиц Красноярского края"». Вступил в силу 1 июля 2010 г.. Опубликован: "Ведомости высших органов государственной власти Красноярского края", №33(404), 5 июля 2010 г. (Legislative Assembly of Krasnoyarsk Krai. Law #10-4765 of June 10, 2010 *On the Registry of the Administrative-Territorial Units and the Territorial Units of Krasnoyarsk Krai*, as amended by the Law #13-6663 of December 1, 2011 *On Changing the Status of the Work Settlements and on Amending the Krai Law "On the Registry of the Administrative-Territorial Units and the Territorial Units of Krasnoyarsk Krai"*. Effective as of July 1, 2010.).
- Законодательное собрание Красноярского края. Закон №13-2925 от 28 января 2005 г. «Об установлении границ и наделении соответствующим статусом муниципального образования Туруханский район и находящихся в его границах иных муниципальных образований», в ред. Закона №9-3727 от 8 октября 2009 г. (в ред. от 12 ноября 2009 г.) «О внесении изменений в Закон края "Об установлении границ и наделении соответствующим статусом муниципального образования Туруханский район и находящихся в его границах иных муниципальных образований"». Вступил в силу через десять дней после официального опубликования. Опубликован: "Ведомости высших органов государственной власти Красноярского края", №5, 7 февраля 2005 г. (Legislative Assembly of Krasnoyarsk Krai. Law #13-2925 of January 28, 2005 *On Establishing the Borders and Granting an Appropriate Status to the Municipal Formation of Turukhansky District and to Other Municipal Formations Within Its Borders*, as amended by the Law #9-3727 of October 8, 2009 (rev. of November 12, 2009) *On Amending the Krai Law "On Establishing the Borders and Granting an Appropriate Status to the Municipal Formation of Turukhansky District and to Other Municipal Formations Within Its Borders"*. Effective as of ten days after the official publication.).
- Туруханский районный Совет депутатов. Решение №7-03 от 10 июня 1997 г. « Устав муниципального образования Туруханский район Красноярского края (http://www.admtr.ru/tr/ystav/)», в ред. Решения №21-523 от 2 июля 2009 г. (Turukhansky District Council of Deputies. Decision #7-03 of June 10, 1997 *Charter of the Municipal Formation of Turukhansky District of Krasnoyarsk Krai* (http://translate.google.com/translate?hl=en&sl=ru&tl=en&u=http://www.admtr.ru/tr/ystav/), as amended by the Decision #21-523 of July 2, 2009.).

Russian_Census_(2010)

The **Russian Census of 2010** (Russian: Всероссийская пе́репись населе́ния 2010 го́да) is the first census of the Russian Federation population since 2002 and the second in its post-Soviet history. Preparations for the census began in 2007 and it took place between October 14 and October 25.[1]

Inside a census station in Severodvinsk

A 10-ruble coin commemorating the 2010 Census

The census

The census was originally scheduled for October 2010, but was moved to 2013 allegedly for financial reasons,[2] although it was also speculated that political motives were influential in the decision. However in late 2009 Russian Prime Minister Vladimir Putin announced that the Government of Russia allocated 10.5 billion rubles in order to conduct the census as originally scheduled (in October 2010).[3] [4]

In the census, many Russians in Siberia entered "Siberian" as their nationality, which prompted Michael Bohm of The Moscow Times to call for removal of 'nationality' from Russian passports.[] 6,000 individuals identified themselves as Siberian in the census.[5]

Preliminary results

Preliminary results released in April 2011 showed the population to stand at 142,905,200,[6] a decline of 2.6% since the 2002 census.

See also

- Demographics of Russia

References

[1] "Осенний пересчет" (http://www.rg.ru/2010/10/15/perepis.html). "Rossiyskaya Gazeta", October 15, 2010

[2] Всероссийская перепись населения переносится на 2013 год (http://www.regnum.ru/news/1206699.html) (Russian)

[3] Путин снова перенес Всероссийскую перепись населения на 2010 год? (http://www.regnum.ru/news/1220355.html) (Russian)

[4] О ВСЕРОССИЙСКОЙ ПЕРЕПИСИ НАСЕЛЕНИЯ 2010 ГО (http://www.gks.ru/perepis/2010/new-perep10.htm) Rosstat Retrieved on November 5, 2009 (Russian)

[5] http://ajw.asahi.com/article/globe/feature/trans-siberian/AJ201209090012

[6] 2010 Russian Census (http://www.perepis-2010.ru/message-rosstat.php) Retrieved on May 25, 2011

External links

- Official website of the 2010 Census (http://www.perepis-2010.ru/) (**Russian**)

Article Sources and Contributors

Evenkiysky_District *Source*: http://en.wikipedia.org/w/index.php?title=Evenkiysky_District *Contributors*: Axeman89, Backspace, Chanheigeorge, Ezhiki, Khanson, Otvis, Paukrus, Volkovtravel, 3 anonymous edits

Krasnoyarsk_Krai *Source*: http://en.wikipedia.org/w/index.php?title=Krasnoyarsk_Krai *Contributors*: Abhijitsathe, Alexbouditsky, Altenmann, Amire80, Anandks007, Anclation, Angusmclellan, Anyep, Apoivre, Articulatron, Avalyn, Backspace, Barend, Bart l, Beagel, Boojum, Bubba Zanetty, CRKingston, CarolSpears, Chanheigeorge, CharlesMoscow, Chumwa, CommonsDelinker, Danny, Dewritech, Ed Cormany, Egon, En.ianm1121, Ezhiki, Foobaz, Good Olfactory, Google Child, Ground Zero, HarDNox, Hectorian, Hellerick, Hibernian, Hottentot, Hunnjazal, IZAK, J04n, JHunterJ, Jesse Viviano, Jetstreamer, Kiril Simeonovski, Kwamikagami, Lady Pablo, Lightmouse, Lugnuts, Mani1, Mattisse, Maximaximax, Mbakkel2, Melnikov, Merovingian, Metricmike, Mintleaf, Mohonu, Morwen, Nanobear, Nightstallion, Nikai, Nikthestunned, OlEnglish, Olivier, Paintman, Pandukht, Paul A, Paulalexdij, Peterlin, Quebec99, Ratzer, Rdfr, Rep07, Rich Farmbrough, Rif Winfield, Roberta F., Russavia, Rydel, Sardanaphalus, Schekinov Alexey Victorovich, Seb az86556, ShelfSkewed, Shoesfullofdust, Skybon, Sluzzelin, Solid dylan, Template namespace initialisation script, Therexbanner, Tibetan Prayer, Tobias Conradi, Topbanana, Velho, Vladpan, Volkovtravel, WolfmanSF, Woohookitty, Work permit, XtalMag, Yannzgob, Ymblanter, Ђыь, , , 39 anonymous edits

Russia *Source*: http://en.wikipedia.org/w/index.php?title=Russia *Contributors*: 'Ndrangheta, *drew, -- April, .:Ajvol:., .mdk., 08OceanBeach SD, 12 Noon, 1234r00t, 1297, 172, 1brettsnyder, 1exec1, 212.192.249.xxx, 24630, 25 Cents FC, 2812, 31ff3y, 32167, 334a, 6 million jews, 66 red hat 66, 747fzx, A Werewolf, A bit iffy, A man without a country, A-giau, A.Savin, A.h. king, A1916, A3r0, AOEU, ARC Gritt, ARUenergy, ASDFGH, Aa75253, Aaker, Aardark, Aaron Einstein, Aaron danielg, AaronPaige, Aaroncrick, Abbadonnergal, Abbeyroad55343, Abce2, Abdominator, Aborig, Abune, Academic Challenger, Accurizer, Ace of Raves, Acer, Acerperi, Acers12, Acio83, Acs4b, AdRock, Adam Carr, Adam78, AdamFouracre, Adambiswanger1, Adashiel, Adeist, Adieujuste, AdjustShift, AdrianTM, Aec is away, Aecis, Aero145, Aesopos, Aeusoes1, Afinogenoff, Afrasclient, Ageekgal, Agent 78787, Agent452, Agomulka, Ahh im going bald, Ahoerstemeier, Ahuskay, Aibur, Aillema, Airplaneman, Aivazovsky, Ajmint, Ajraddatz, Akanemoto, Akbg, Akhristov, Akshaysrinivasan, Al3xil, Alaexis, Alam82, Alaney2k, Alangstone, Alasdairking, Albanaco, Aleenf1, Aleksei, Alethiophile, Alex '05, Alex Bakharev, Alex earlier account, Alex756, AlexNebraska, AlexPU, Alexander Domanda, Alexander Gerashchenko, AlexanderKaras, Alexikoua, AlexiusHoratius, Alexnevzorov, Ali, Aliyevramin, Allan Letinov, AllenHansen, Alloy, Allysia, Alnrr, Alphachimp, Alphasinus, Altau, Altenmann, Alton, Alynna Kasmira, Amberrock, Ameliorate!, Amenzix, Amire80, Amrad, Amylyn, Anandks007, Anddrex, Andonic, Andonio, Andre Engels, Andres, AndrewDressel, AndrewHowse, Andrijko Z., Andris, AndriyK, Android79, Androidsdungeon, Andy Marchbanks, Andy120290, Andyjsmith, AndytheTweedale, Ang15, Angelo De La Paz, Anger22, Angr, Angusmclellan, Anonymous from the 21st century, Anonymous from the 21th century, Anonymous44, Answerthis, Antandrus, Anthonyd3ca, Antipode, Anton Fischer, Anton Khorev, Antonio Basto, Anubis-SG, Anvyder, Aogouguo, Aphaia, Apoivre, Appleboy, Aprolix, Aramgutang, Arbor to SJ, Arcenciel, Archer3, Archiver of Records, Arem04667, Ares1991, Aria613, Arichnad, Arigato1, Armanos, Armydepot, Arne List, Aromatkzn, Art LaPella, Arthur889, Article editor, Arturo313, Arun11, Asdfking1, Ash sul, Asidemes, Assassin3577, Assyrial, Astronaut, Astropithicus, Astrotrain, Atilano, AtilimGunesBaydin, Atitarev, Atlantas, Atoric, Aude, AugustinMa, Auntof6, Ausir, Avala, Avaya1, AvicAWB, Avicennasis, Awild4, Awild5, AxG, Axegod12, AxelBoldt, Axt, Ayle001, Ayrton Prost, Azov, AzseicsoK, BD2412, BGManofID, Bachrach44, Backspace, BadHabit86, Bahavd Gita, Bakerq, Baldhur, Ballestrain, Bambuway, BanyanTree, Barbe Igor, Baristarim, Baronnet, Bart133, Basawala, Bazonka, Bbryce1, Beach Coma, Beagel, Beatle Fab Four, Beaumain, Beeblebrox, Beenhj, Beeswaxcandle, Beetstra, Bejnar, Belligero, Ben-Velvel, Bender235, Bendono, Benton287, Benzach22, Beregond5, Berkunt, Bernard the Varanid, BernardTom, Bert Schlossberg, Bertie, Bertyrex, Bestalex, Bestsemesterinfo, Beyond silence, Bezenek, Bhadani, Bhargav123, Bhb1991, Bibliomaniac15, Big Moira, Big iron, Biggerj1, Bigthu77, Bility, BillSMASH, Billlion, Billybill, Billybob68, Billybobfro, Binokoron, BirdValiant, Birlanady, Biruitorul, Bjelleklang, Bjmarfito, Bkell, Bkonrad, Bkwillwm, Blackjack48, Blamethrower249, Blaylockjam10, Blethering Scot, Blindman shady, Blizzardstep0, Blood Red Sandman, Bloodfinger, Blue william, Bluecheeseboy, Blueconno, Bluehat66, Bmicomp, Bnwwf91, Bob rulz, Bob999991, BobDively, Bobanni, Bobblewik, Bobianite, Bobo192, Bobrayner, Boenrace, Bogdan, Bogdangiusca, Boing! said Zebedee, Bolonium, Bonadea, Bonde, Bongwarrior, Bonheir, Borisblue, Bornhj, Borovv, Borsi112, Bosonic dressing, Bowei Huang 2, Br'er Rabbit, Bradish, Brandmeister, Brandmeister (old), BrendelSignature, Brewmanz, Brian0918, Brianski, Brichcja, Bridgetansel, BrigadierGeneralBaptisms, Brion VIBBER, BritishWatcher, Brutaldeluxe, Bry9000, Bsdlogical, Btmachine333667, Buaidh, BubbleDude22, Buckshot06, Buffadren, BuickCenturyDriver, Bulatych, Bull Market, Bulldog73, Burbrack, Burningliquid, Bushy moustache, Butros, Buttons, Butwhatdoiknow, Bvjrm, C filev, C-b90, C12H22O11, CJ King, CRKingston, CTJames, CVTautomaton, Cactus.man, CalJW, Calbob66, Calliopejen1, Callymoto, Calmer Waters, Calton, CambridgeBayWeather, Camembert, Camw, Can't sleep, clown will eat me, Canadian-Bacon, CanadianLinuxUser, Canderson7, Caniago, CanisRufus, Canley, Canterbury Tail, Cantus, Capitalistroadster, Captain, CaptainFugu, Caratinga8, CaritasUbi, Carlodellora, Carn, Carnby, Casliber, Cassini83, Catgut, Ccarroll, Cccccccccc, Cchow2, Cdc, Cedrus-Libani, CeeWhy2, Celestra, Certh, Cezzvil, Cfeet77, Cgoodell, ChKa, Chairboy, Chairman S., Chandler Fleck, Chaplin62, Char6128, Charles Essie, CharlesMoscow, Charliemeyers, Chase me ladies, I'm the Cavalry, Chauncyo9, Chavash, Che829, Check two you, Cheese007, Chemical echo, Cherkash, Cherri65, Chochopk, Choop, Chowbok, Chris Roy, Chris the speller, Chris53516, ChrisO, Chrisconey, Chrisfarley1sfat, Chrishmt0423, Chrislk02, Chrism, Chrisminter, Christ2death, Christian List, Christopher Parham, Chtovimenitebemojem, Chun-hian, Chunky Rice, Chzz, Cienade, Cirt, City of Destruction, Civil Engineer III, Ckatz, Claidheamohmor, Clarince63, Clawed, Clcsfist, Cmapm, Cms479, Cod1337, CodeMonk, Codetiger, Coelacan, Colchicum, ColdWarCharlie, Coldsnort, Colipon, Coloane, Colonies Chris, Commander Sergei Bjarkhov, Commi9000, CommonsDelinker, Comp25, Computerjoe, Conman666, Connormah, Conscious, Constantijn, Constantzeanu, Conte di Cavour, Conversion script, Cooldude7273, Cor1314, Coredesat, Corticopia, Cottelli, Cousinjamby, Cptnono, Crazycomputers, Credema, Cremepuff222, Crism, Crit222, Croc22, Crocodealer, Croperz, Crotalus horridus, Cruzian, Cs-wolves, CsDix, Curps, Curriec69, CyberSkull, Cyblue, CyclePat, Cyclonenim, Cylik, CyrilB, Cyrius, CzarB, D-star22, D0t, D4niel11, D6, DARTH SIDIOUS 2, DCGeist, DCowhig, DDima, DJ Clayworth, DO'Neil, DTOx, DVD R W, DVoit, Daa89563, Dabigjc, Daborhe, Dacrone, Dadaist6174, Daduzi, DaemonischEngel, Dalf, Dalta, Damien solta, Daminstant, Dan D. Ric, Dan56, DanKeshet, Dancer6, Daniel5127, DanielCD, Danielkueh, Danieltiger45, Daniil naumoff, Danio, Bibliophylax, Danlaycock, Danlina, Danny, Dantheman531, Dar-Ape, Daraheni, Daremyth, Dark Shikari, DarkElrad, Darkmasterjoey, Darknshadow, Darkspots, Darkwarden, Darkweasel94, Darrendeng, Darthgriz98, Darwinek, Davandron, Davemcarlson, Davenbelle, Daveswagon, David Falcon, David Gerard, David Johnson, David Jordan, David Kernow, David.Mestel, Dbachmann, Dbenbenn, Dcooper, DeServe77, Deacon of Pndapetzim, Deavenger, Debresser, Decman800, Def43, Deflective, Deg2800, Delirium, Delldot, Delxx, Demerzel, Demiurge, Demmy, Dendodge, DenisLG, DenisRS, Denniss, Deon, Der Golem, Descendall, Desiphral, Detnchris14, Deviathan, Dewritech, Dfhjhjjftksfbgr, Dhruvgupta97, Dialog, Didactohedron, Dief88, Diegom809, DigitalGhost, Digwuren, DimaY2K, Dimmy Simmy, Dina, Dino, Dirak, DirkvdM, Discospinster, Djcartwright, Djr13, Dk1panther, Dlrohrer2003, Dlyons493, Dmerthe, Dmeyler, Dmismir, Dmitri, Dmitri Lytov, DmitryKo, Dmn, DoRD, Doc glasgow, DocSigma, DocWatson42, Docu, Documentinghalloween, Dodiad, Dojarca, Dominik92, Donut2, Doopdoop, Dori, Doubleodd, Doug Johnson, DougRWms, Download, Downstrike, Dowsiewuwu, Dppowell, Dputig07, DrJamesMarcus, Draeco, Dralwik, Dramartistic, Drappel, Drbug, Dren55, DriveMySol, Dropbows90, Dru of Id, Dsazonov, Dsp13, Dudtz, Duncharris, Duroy, Dustimagic, Dwarf Kirlston, Dwayne, Dwo, Dwrcan, Dyadyavasya, Dycedarg, Dysepsion, Dysprosia, Dzenanz, Dzied Bulbash, Dÿrlegur, E Pluribus Anthony, E0steven, EJF, ESkog, Eaefremov, Eagle4000, Easybreezy1023, Eazymofuzer, Ecad93, EconomistBR, EddyVadim, Eden87, Editor88, EdoDodo, Edoroth, Edward, EdwardLane, Eewild, Efghij, Ehsan1981, Eightofnine, Eisnel, Ekren, El C, El Snubbe, El Suizo, ElKevbo, Electionworld, Elementanton, Elk Salmon, Elliegurl11, Elmo70, Elockid, Eloquence, Elostirion, Emax, Emberstone, Emery2983, Emigrant123, Emilfarb, Emoelvis, EncephalonSeven, EncycloPetey, Ender78, EngGerm12, Enkyo2, Enormousdude, Enviroboy, Epbr123, EpiC, Epicbobman10, Epicgenius, Eric-Wester, Erik9, Esimal, Esn, Estoy Aquí, Estropes, Eszett, Ethanisasbeancool, Ethelred Cyning, Eu.stefan, Eugrus, Euoa, Eurocopter, Evandroroncatto, Evangelion9490, Everton, Everyking, Evil Monkey, Evilhairyhamster, Ewlyahoocom, Excirial, Explicit, Ezhiki, Ezra Wax, FCSundae, FF2010, Facts707, Fair Deal, Faithlessthewonderboy, Fakirbakir, Falcon8765, Fame, Fan-1967, Fang Aili, Fasgab, Fatal exception, Fatslob, Fedallah, Feedmecereal, FeelSunny, Feeltherbbrducki, Feezo, Felipe Menegaz, FellGleaming, Fellix, Fernirm, Feydey, FidelFair, Fieldday-sunday, Fil usa, Filipdr, Finlay, Fireaxe888, Firsak, Firsfron, Fiscodrawing, FisherQueen, Fisss, Fist of Glory, Fitnr, Fiveyearplan, Flakybizcuit, Flamarande, Flashflash;, Flatterworld, Flexx91872, Flix11, Flowonthego, Flrn, Flubeca, Flying Llamas, Folantin, Football1305, Fram, Frances craig, Francis Tyers, FrancisTyers, Frank Russian, Freakofnurture, Freddie42, Freddiem, Fredrik, Freivolk, FreplySpang, FreshBulletTime, FreshFruitsRule, Fresher, FreyasCrystalizedAngels, Freyr, Frietjes, Frito-lay12345, Frostbitn, Funandtrvl, Funnybunny, Funnyhat, Furrykef, Future Perfect at Sunrise, Fvw, Fys, G. Campbell, G8summit, GHe, Gabbe, Gaius Octavius Princeps, Galati, Galaxydog2000, Galoubet, Gamalon Idogyr, Gamingexpert, Gandygatt, Garik, Garret Beaumain, Garth of NEaB, Gary D, Garzo, Gavrilov, Gaylord22, Gaz, Gdo01, Gearfordeath, Gelovotoz, Gene s, General Eisenhower, General Grievous, Geni, Geoffrie, Geoffspear, Geohem, GeorgeLouis, GerardM, Gertjan R., Gggh, Ghirlandajo, Gibson Flying V, Gidonb, Gigety gigety goo, Gio11211, Girdi, Gjd001, GlaDooo, Glen, Glenn, Globalstatus, Glossologist, Gnomz007, GoOdCoNtEnT, Gogo Dodo, Gohho, GoingBatty, Golbez, Gold heart, Golfcourseairhorn, Good Olfactory, GordyB, Gosso, Gostanford8, Gr8opinionater, GraemeL, Grafen, Graham87, GrahamColm, Grapeme, Graphic, Green Giant, GreenUpGreenOut, Greenshed, GregAsche, GregorB, Gregorik, GrelkJem15, Grendelkhan, Greswik, Grey Wanderer, Greyhood, Griffinofwales, Gringo300, Grosplant, Ground Zero, Grunt, Grye, Gryffindor, Grúnt, Gscshoyru, Guilherme Paula, Gump Stump, Guppie, Guy Peters, Gwdr500, Gzornenplatz, H-ko, HIDECCHI001, HJ32, HJV, Haaqfun, Hadal, Hadiyazzi, Hadseys, Hagedis, Hainzey, Hajor, Half price, Halmstad, HammerHeadHuman, Hanacy, HansHermans, HanzoHattori, Happyme22, Hari6389, Harmil, Harp, Harryboyles, Harveyqs, HasanFerdinand, Hazhk, HaŋaRoa, Hbdragon88, Hdt83, Heff01, Hellerick, Hello32020, HelloAll12, Hellogeoff, Helsinkicommission, Hemanshu, Henk65, Henry W. Schmitt, HenryLi, Heron, Hertok, Herut, Hfsulliv, Highfields, Highvale, Hirudo, HisSpaceResearch, Hisagi, Hiyaninja, Hmains, Hmusseau, HobbesLeviathan, Hodja Nasreddin, HoldenV8, Hollowpens, Holod, Homagetocatalonia, Homero simpstone, Hongshi, Honshuzen, Horlo, Horn123, Horvat Den, HotXRock, Hottentot, Howcheng, Hto9950, Hullaballoo Wolfowitz, Hungaria777, Husond, Hut 8.5, Huw Powell, Hvn0413, Hydriz, Hydrogen Iodide, Hz.tiang, I63G, IGod, IJzeren Jan, IRelayer, IVRacing, Ialsoagree, Iampatel, Ian1981, Ianneub, Iatomm, Icairns, IcePuckScore, Icseaturtles, Idaltu, Idet.proverka, Ief, Ienjoyeating, Ifeldman84, Ifnord, Ifyoudontlikethisthenpleasedonttellme, Igiffin, Ignatpakhotin, IgorMagic, Igordebraga, Iheartjesusveryverymuch, Iketsi, Ikh, Ikiroid, Iksisy, Illegitimate Barrister, Illythr, Iluvjazz, Ilya1166, IlyaMart, Ilyusha V. Novikov, Ilyushka88, Impala2009, Imperatorjan, Improv, Incidious, Indisciplined, Indon, IngSoc BigBrother, Inglok, Ingolfson, Inhakito, Innab, Insuranze, Int21h, Intelligentsium, Inter, Intothewoods29, Introvert, InvisibleK, Inwind, Ioakinf, Ioannes Tzimiskes, Ionut.shaggy, Iridescent, IronPhoenix, Irpen, Irrypride, Isaac Rabinovitch, Ischorr, Islamablad, Istartfires, Italiano111, Itinerant1, Iv1607, Ivan Volodin, Ivan2007, Ivanip, Ixfd64, Izax143, Izzedine, J Di, J Milburn, J budissin, J.K Rowling, J.delanoy, JD79, JForget, JHMM13, JIP, JJARichardson, JNW, JRHorse, JSquish, JTBX, JWSchmidt, Jacek Kendysz, Jack Silver1000, Jack Upland, Jackfield, JackofDiamonds1, JackofOz, Jacksaywhat, Jacoplane, Jafeluv, Jagged 85, Jam1993, JamesBWatson, Jamesday, Jamesinderbyshire, Jamesooders, JamieS93, Jankaspar, Janneman, Janviermichelle, Jarble, Jarkeld, Jaronnelson, Jason Recliner, Esq., Jatrius, Java13690, Jaxl, Jaydlewis, Jayron32, Jbl1970, Jcaragonv, Jcrook1987, JdeJ, Jdforrester, Jebba, Jeendan, JegaPRIME, Jennavecia, Jennica, Jensboot, Jerome Charles Potts, JerryFriedman, Jerzy, Jevansen, Jez, Jezhotwells, Jh51681, Jhendin, Jiang, Jim Douglas, Jimp, Jimtaip, Jirka.h23, Jj137, Jjhcap99, Jjkkl, Jkaplan, Jniech, JoSePh, Joachim Weckermann, JoanneB, JodyB, JoeSmack, Joefu, Johan Elisson, John, John Price, John Quincy Adding Machine, John Smith M.D., Ph.D, John of Reading, John254, JohnG62, JohnWittle, Johndyra, Johnhenry312, Johnleemk, Johnluisocasio, JohnnyWiki, Jojolias, Jomorepinch, Jonatan Swift, JonathanDP81, Jonathunder, Jonearles, Jonjames1986, Jonkerz, Jose77, Joseph Solis in Australia,

Joshualoljosh, Jossi, Jovianeye, Joy, Joyous!, Jozdnog, Jpeob, Jpgordon, Jph, Jpople, Jrockley, Jtkiefer, Jugbo, Julekmen, Jumbuck, Jung dalglish, Jusdafax, Jusjih, Just zis Guy, you know?, Justinjuicebox, Jvanore68, Jwissick, K a r n a, K.O.T., KNewman, KTC, Kablammo, Kaiser matias, Kajervi, Kalsrag, Kanchelskis-it, KapilTagore, Karabinier, Karesu12340, Karl Marx, Kascnef82, Kassabov, Katydidit, Kayac1971, Kaz, Kazak, Kazayta, Kbh3rd, Kd4nuh, Kda2011, Keegan, Keelm, Kefeer, Keilana, KeithB, KelleyCook, Kelstonian, Keluar94, Kelvinc, Kemor, Kenmore, Kev2310, Kevin B12, Kgrad, Khashyar, Khatru2, Khazar2, Khelnor, Khoikhoi, Khvalamde, Kids4Fun, Kikichugirl, Kilhan, Kilimac, KillaShark, King Lopez, King Zebu, Kingpenguins2007, Kingpin13, Kintetsubuffalo, Kkkdc, Kkm010, Klenje, Kman665, Kmorozov, Kneiphof, Knutux, Koavf, Kolja21, Kolkhian, Koov, Kosebamse, Kotniski, Koyaanis Qatsi, Kozuch, Kpalion, KrakatoaKatie, Kralizec!, Krawndawg, KrayCeremoush, Krenakarore, Krich, Kristallstadt, Kross, Krukowski, Ktoto89, Kuban kazak, Kubigulo, Kuiavian, Kuimov, Kukini, Kulikovsky, Kungfuadam, Kungming2, Kurieeto, Kurtber, Kuru, Kurykh, Kusma, Kusunose, Kuzmaxster, Kwamikagami, Kwertii, Kwesadilo, Kyle Barbour, L Kensington, LEGOBRICKSTER3, LMANSH, LUCPOL, La goutte de pluie, Lacrimosus, Lactasamir, LadnaV, LahDeeDah7, Laikas, LairepoNite, Lakers, Lantern40, Lappip, Lapsed Pacifist, Le Anh-Huy, Leadgold, Leandrod, Leapsorymt, Lear's Fool, Lectonar, Lenev, Lennonist, Leolaursen, Levelistchampion, Levineps, Lexa87, Lexicon, Liamdaly620, Library time, Life is like a box of chocolates, Liftarn, Lightdarkness, Lightmouse, Ligulem, Likelife, Likthem, Lilac Soul, Lilmaniac, Limideen, LincolnSt, Lindenhurst Liberty, Ling.Nut3, Linguist Dan, Linkfan, Linnell, LiquidOcelot24, Little Professor, Livajo, Llamabread, LlywelynII, Lockesdonkey, LokiiT, Lokqs, LonelyMarble, Longhair, LookingGlass, Loonymonkey, Looxix, Lord Vader, Loren36, Lothar von Richthofen, Lotje, Lowe4091, LtPowers, Luisaterzontz, Luisztdt, Luk, Lumen24, Luna Santin, Lunchboxhero, Lupo, Lvivske, Lysy, M.nelson, MC MasterChef, MER-C, MJCdetroit, ML, MONGO, MSGJ, MZMcBride, Mac, Macintosh User, MackSalmon, Mad Max, Maelnuneb, Magicmonster, Magister Mathematicae, MagnusA, Mailer diablo, Majorly, Mak2007, MakmoudHassan, Malarious, Malcolm Farmer, Male1979, Malerin, Malhonen, Malick78, Malmo Blue, Malo, Man vyi, Mangalaiii, Mani1, Manop, Mansmokingacigar, MapsMan, Maravjua, Marc87, Marcika, Marco Neves, Marcus2, Mardus, Marek69, Mareklug, Mariapoliantseva, Mark rico, MarkGallagher, MarkSweep, Marko7, Marlow4, MarphyBlack, MarsRover, Martarius, Martin23230, Martin451, Martinwilke1980, MasterShake1000, Materialscientist, MathMartin, Matreal, Mattbrundage, Mattisse, Mattrulez, Matts computer, Matty, Maunus, Mausy5043, Mav, MaxSem, Maxim, Mazerunner, Maziotis, Mb nl, McDogm, McGeddon, Mcorazao, Meadow Soprano Wheels!, Mediolanum, Medo9, Medvedev40, Meegs, Mefistofele, Meganhernandez, Melikamp, Mememethem, Mentifisto, Merbabu, Merenzon, Merotoker1, Mesgul82, Meursault2004, Mfa fariz, Mfzb04, Mgiganteus1, Mhacdebhandia, Mhardcastle, Mherlihy, Mhocker, Mhsb, Mic, Michael Devore, Michael Frind, Michael Hardy, Michael5046, MichaelMaggs, Michaelas10, Michaelmas1957, Michaelphelps'sbongwater, Michaelwuzthere, Mike Rosoft, Mike by, Mike1024, Mikhail Garouznov, Milkyeyes, Miller17CU94, Mimihitam, Minesweeper, Minimac, Mirv, Misaq Rabab, Misortie, Missionary, Mistakefinder, Misza13, Mitchking23, Mitzu17, Miyokan, Mjpieters, Mobile6681, Modulatum, Modþryð, Moejoeonline, Mohawkjohn, Molly-in-md, Monedula, Monk, Monkeyman, Monterey Bay, Mooo, Moravice, Moreschi, Morgankevinj huggle, Morwen, Moscvitch, Moshe Constantine Hassan Al-Silverburg, Moxfyre, Moxy, Mqe, Mr Bartels, Mr Stephen, Mr. Lefty, Mr. Military, Mr. Yooper, Mr.Unknown19, MrJacky, MrSchmenge, Mrbrush, Mrs Trellis, Mrtwig, Ms2ger, Mschel, Mschiffler, Muhaidib, Muncadunc, Munci, Muntuwandi, Musaabdulrashid, Mushroom, Musical Linguist, Musser, Mustafagheran, Mwanner, MyNameIsNotBob, Myanw, Myunknownside000, Mzajac, N-edits, N-true, N3X15, NHRinger, NIR-Warrior, NPPyzixBlan, Naddy, Nadirali, Nadyes, Nakon, Nanobear, Nanshu, Narking, Narxysus, Naryathegreat, Nat Krause, Natalia O., Natalie Erin, Natgoo, Nathanww, NativeForeigner, Natl1, Naturada137, NawlinWiki, Nayvik, Neartobrum, Neegjeris, Neglen, Nehrams2020, Neil916, Neonmario, NerdyScienceDude, Netoholic, Netol, Neumannk, Neutrality, NeutronTaste, Never give in, NewEnglandYankee, Newsrussia, Ngyikp, Niceguyedc, Nichalp, Nick Number, Nick mallory, Nick-D, Nick.mon, Nick125, Nickshanks, Nicksss93, Nicksukh, Nickx33, Nickydude, Nicolas Love, Nigholith, Nightstallion, Nihil novi, Nihiltres, Nikai, Nikitadanilov, Nikkimaria, Nikola Smolenski, Nirvana888, Nissi Kim, Nivix, Nixer, Nizzil, Njaelkies Lea, Njeman, Njoedits, Nlight2, Nmpenguin, No Guru, No limits99, No.13, Noah Salzman, Nobleeagle, Noiseidea123, Noisettes, Nojustno, Nono64, Noodlenut, Noommos, Noozgroop, NorthernThunder, Northumbrian, Notedgrant, Notheruser, Notthe600, Nsk92, NuclearVacuum, NuclearWarfare, Nug, Numbo3, Nv8200p, Nyttend, Nérostrateur, OBdude7, Obradovic Goran, Observerq, Occamy, Octane, Ogatai, Oghmoir, Ogmo, Ohconfucius, Ohmygod766, Ohnoitsjamie, Okapi, Okolobaxa, Okome, Olegthezenon, Olegwiki, Olgasavina, Olgerd, Oliphaunt, Olivier, Ollivier, Olna, Olorin28, Olvegg, Omega360, Omicronpersei8, Ominpanda, OneGuy, Onehundredyears, Only Red, Onopearls, Onorem, Openrussia, Opodo123, Optigan, Orangerider, Orenburg1, Orionist, Orthuberra, Ostap R, Otolemur crassicaudatus, Ouip, OverlordQ, Overnight darkness, OwenBlacker, OwenX, Owned4itall, Oxana879, PANONIAN, PFHLai, PIO, PRH, PSzalapski, PTYLER22, Pablo X, PaddyBriggs, Paine Ellsworth, Pajfarmor, Pakhomovru, Palefire, Pama73, Panairjdde, Papa Carlo, Parhamr, Parishan, ParisianBlade, Pass a Method, PasswordUsername, Pathoschild, Patrick, Patrick Berry, Paukrus, Paul August, Paul Benjamin Austin, PaulGarner, PaulHanson, Paulinho28, Paulscho, Pavel Vozenilek, Pax, Pax:Vobiscum, Pb30, Pearle, Pedroshin, Pegasus1138, Pensionero, Penubag, Perceval, Pereleshin, Permico, Perovic, Persian Poet Gal, Peru-Russian888, Peruvianllama, Peter Horn, Peter Isotalo, PeterPredator, PeterisP, Pethr, Petri Krohn, Peyman Ghasemi, Pgan002, Pgk, Phaedriel, Philip Trueman, Philipmj24, Phillip J, Philpm930, Pholi, PiMaster3, Pianist ru, Picaroon, PickupmanUP1966, Piepoop, Pigman, Pilcrow, Pilotguy, Pinball22, PinchasC, Pinethicket, Piotr9, Piotrus, Pip92, Piratedan, PizzaBox, Plapre09, PlatypeanArchcow, Pletet, Pmanderson, Poetaris, Poker face20, Polaron, Polly, Polmandc, Polnian, Polon11, Polsequ95, Poor Yorick, Poptartbeebee, Porkies84, Postdlf, Pouya, Pras, Prawn3944, Princesin72, Principia1, Profmajom, Pronoun, Prospectmira, ProtectionGate, Pseudomonas, Psy guy, Ptorchilo, Ptrt, Pudeo, Purpleturple, Pygenot, Pyroclastic, Python eggs, Pz821, Q43, QFlux, QZXA2, Qasaqsuyu, Qatter, Quandapanda, Quarkington, QueenCake, Queenmomcat, Qutezuce, Quuxplusone, Qwertzy, Qxz, R'n'B, R-41, R7K, RJN, RK, ROSARIO STEPHEN, RPIRED, Rab V, Rabindra Baral, RadicalBender, RadioKirk, RafaAzevedo, Rahlgd, Rail88, RainWingWolf, Raja Hussain, Ral315, Ramand, RamblinWreck3, Ramdrake, Rameriz, Ran, Random contributor, RandomP, Randomblue, Ranveig, Rarelibra, RaseaC, RashersTierney, Rasmus Faber, RasputinAXP, Raven4x4x, Raymond arritt, Razionale, Rclark53, Rd232, Really Spooky, Recker, Reconsider the static, Red Director, Reddi, Reddragon555632, Redking7, Redstarsldr, Redvers, Regionalsimp, Rehnn83, Reichenbach, Remisc, RenamedUser01302013, Reticulated, Retired username, RetiredWikipedian789, Rettetast, Rev4ricky, RexNL, Rezashah4, Reznoff4, Rhysn, Riana, Rich Farmbrough, Richard David Ramsey, Richard0612, RichiH, Rick Block, Rick DeLong, RickK, Ricky81682, Ricky@36, Rickyrab, Rivertorch, Rjcflyer@aol.com, Rje, Rjensen, Rjwilmsi, Rmcubed, RoadTrain, Robert K S, Robert Merkel, RobertG, RobertLunaIII, Robertgreer, Robertp72, Robinkruyt, Robwingfield, RockMFR, Rockador, Rocket71048576, Rodasmith, Roentgenium111, RohirrimRider, Romanius, Romanm, Rome1453, Ronbo76, Rory096, RossPatterson, Rossen3, Rowe 13, Roy da Vinci, RoyBoy, Royalguard11, Rrburke, Rreagan007, Rrtycoon, Ruhrjung, RuineR, Runningfridgesrule, Ruok, Rusish, Russavia, Russian Luxembourger, Russian-Reaper, Russiansoul, Rwxrwxrwx, RxS, RyanGerbil10, Ryanforesight, Ryulong, S.B. Odin, S119234, S5switch, SCJohnson77, SD6-Agent, SDC, SECProto, SGBailey, SGGH, SJK, SMC, SOPHIAN, SPORTYGRL012, SQGibbon, SS451, STARWARSGEEK, SWO1975, Saffloped, Sagaciousuk, Saganaga, Saiga12, Saluyot, Sam Korn, Sam Weber, Samantha555, SamdupLa, SamdupLama, Samir, Samrolken, Samuel Blanning, Samuelsen, Samwb123, Samy511, Sanbec, Sandip90, Sandlas Juagas, Sandstein, SandyGeorgia, Sango123, Sardanaphalus, Sarranduin, Sasa2007, SashaGolikov, Sasquatch, SatnSpwn, Satori Son, Sayden, Sca, SchfiftyThree, SchmuckyTheCat, Schnellundleicht, SchnitzelMannGreek, Schooks59, Schoolthing, SchuminWeb, Scientistmohamed, Scipius, Sciurinæ, Scott Burley, Scottwilleke, Scrazen, Scythian1, SeNeKa, Sean K, Secfan, Seicer, SelfQ, Selfworm, Seniorfox, Senzangakhona, Sergey.belyakov@gmail.com, Seryo93, Sesu Prime, Seth harsch, Sethmahoney, Sevela.p, Sevenlee, Seyrüsefer, Sfahey, Shadiac, Shadow Android, Shadow1, Shadowjams, Shadowxfox, Shak-74M, Shamir1, Shanes, Shannon bohle, Shattered, Shattered Wikiglass, Shisock, Shiva Evolved, Shizhao, Shoaler, Shogunpk, Shredder2012, Shultz IV, SiPlus, Signalhead, Signsolid, Silly rabbit, SimonP, SimonX, Simonsaysabc123, Sindri, Sir Edgar, Sir Nicholas de Mimsy-Porpington, Sirgregmac, Sirius85, Size J Battery, Sjakkalle, Skjorgen, Sky Attacker, Skybon, Slakr, Slashme, Slif3r, SlimVirgin, Slomox, Slowking Man, Slurpy121, Slyguy, Smallbones, Smalltime0, Smarkflea, Smashing, Smb1001, Smeira, SmthManly, Snaildaddy, Snailwalker, Snap Davies, SndrAndrss, Snowolfd4, SoLando, SoSaysChappy, SoWhy, Sociallyawkwardpenguin, Soetermans, Sokol 92, Solentways, SolidEdgeGuRu, Solitaryxchild, SomamcFloater, SomeFreakOnTheInternet, SomeStranger, Something20, Somethingsometing, Son of thunder, Sonicsuns, Soosed, Sophus Bie, Sour cream monster8, Sozsoz, Sp33dyphil, Spaceboy492, Spangineer, Spartan65116, SpartanG117, Spartaz, SpecMode, Spellcast, Spespatriae, Spettro9, SpigotMap, Spike35031, Spliffy, SpookyMulder, Spot87, Spotty11222, Spring01, SqueakBox, Squeezeweasel, Squiddy, Squish2013, Srushe, Sshirokov, Ssolbergj, Ssr, St.daniel, Stalin88, Stan Shebs, Standfest, Starnestommy, Starwars10, Stassats, Staszek Lem, Staxringold, Stealthbreed, Steel, Steinbach, Stella Vesper, Stephen MUFC, Stephenb, Stephreef, Steve2011, Steven J. Anderson, Stevenmitchell, Stevenochs, Stitchix, Stoljaroff1987, Storkk, Student7, Subtropical-man, Suede, Suffieldlax24, Suffusion of Yellow, Suigntou11011111, Suklaa, Sungjwoo, Sunray, Super-Magician, SuperDeng, Superdude602, Superking, Superninjakatmanofeditorialninjakats, Supertask, Superzohar, Supreme Deliciousness, Surferus, Surlyduff50, Svmich, Sweden555, Swikid, Swimm2, SwinSvinoza, Swu1992, TAG-A-b10, TFCforever, TKD, TUF-KAT, Taamu, Tajik24, TakeMemoriesAway, Tamarav, Tango, Tarret, TarzanASG, Tasc, TastyCakes, Tat1642, Tataral, Tauiris, Tavork, Tavrian, Taw, Tb, Tbhotch, Tealdon, Teamweeeman, Techman224, TedE, Teevek, Teh1337indian, Tennisuser123, Teo jun min, Terailer, Terence, Tesi1700, Testu, TexasAndroid, Texmon, Texture, Tgv8925, Thatswhatsup, The Anome, The Banner, The Epic Hawk, The Green Fish, The Rambling Man, The Thing That Should Not Be, The Tonya, The Transhumanist, The pencil fudge, The tooth, The71, TheAnimalSkull, TheChanger, TheDJ, TheLeopard, TheMadBaron, TheNewPhobia, TheRanger, Thecrazydonvingigante, Theda, Thedcm, Thedefenceman, Thedjatclubrock, Theguyotc, Thejadefalcon, Themightyquill, Thenowhereman, Theprophetmyass, Therequiembellishere, Therexbanner, Theseven7, Thesevenseas, Thryduulf, Thue, Thunderrat34, Tide rolls, TigerShark, Tigerhawk47, Tim Starling, TimBentley, TimComm, Timberlax, TimothyPilgrim, Timrollpickering, Tinsue, Tired time, Tisquantum, Titoxd, Tiwonk, Tjl1128, Tobby72, Toccata quarta, TomStar81, Tombrider1, Tombseye, Tomcat7, Tomjoneser, Tonik, Tony1, Tonyfaull, Topgun205, Topuria, TortoiseWrath, Tovarish14, Toy111, Tpbradbury, Tra, Trademarx, TransUtopian, Transilvanian Dracula, Trasman, Treveraritz, Trevor MacInnis, TrollingTrollerTroll, TrumanRu, Trusilver, Trust Is All You Need, Tsirk13, Tsob, Ttony21, Tulandro, Tulkolahten, Turgeis, Turgidson, Turkeli, Turlo Lomon, Tvdog, Twillisjr, Ty1993, Typhoonchaser, TyrocP, Tyw7, U.Steele, Ubiquinoid, Ucucha, Ufwuct, Ugen64, Ugur Basak, Ulises Heureaux, UltimatePizzaLord, Uncle Dick, Uncle Scrooge, Underlying lk, Underorbit, UnicornTapestry, Unionin, Universe=atom, Unsc, Unschool, Untifler, Unused0024, Urfinze, Uris, Urod, Useight, User BartS, User27091, User6854, User86654, Usergreatpower, Utcursch, Uuu87, V4711, VORON SPb, VVPushkin, Valentinian, Valentino, Valio subaru, Valoradam, Van der Hoorn, Van helsing, Vancouverguy, Vaniba12, Vanished user 05, Vanished user 9i39j3, Vanished user aqpoi4u3tijsrfi, Vardion, Varnav, Vary, Vas123, Vega84, Vegaswikian, Vegidio, Venajaguardian, Ventur, Verdadero, Verseruns, Vetrov, Vice regent, Viciousvogel, Victor12, Victoria2007, Victory in Germany, Vidokov, Virgule82, Vlad Tepes IV, VladPutin, Vladdacomrade, VladimirKorablin, Vlasds, Vlastoff, Vmenkov, Vmrgrsergr, VolatileChemical, Volodin, Volodyanazarkevych, Volunteer Marek, Vortexica, Voyevoda, Vrghs jacob, Vrinan, Vroman, VsevolodKrolikov, Vsmith, Vssun, Vuo, Väterchen, WGee, WLFV, Wafulz, Wallie, Wanjuscha, Ward3001, Warfvinge, Wassamatta, Wassermann, Watashiwabakayo, Waterhoof, Wavelength, Wavesswung99, Wayne317, Wayward, Wecameasromans, Weedro, Welsh, Wenli, Wesha, WesleyDodds, Wetnose23, WhereAmI, Whichtrial, WhisperToMe, WhitmanNE, Who, Wicojrpr, Wiglaf, Wik, WikHead, Wiki alf, Wiki1609, WikiLaurent, Wikiagoo, Wikibojopayne, Wikiborg, Wikieditor06, Wikignome1213, Wild wadi, Wildyoda, Wilfried Derksen, William Allen Simpson, William Avery, Willqualey, Wimt, Windman233, Wisco, Wjejskenewr, Wjfox2005, Wk muriithi, Wknight94, Wliiam·Shakespeare, Wollont, WoodElf, Woohookitty, Worldtimezone, Worldtraveller, Wrath of Hell, Wsiahoffman, Wælgæst wæfre, X!, X96lee15, XAKxRUSx, Xahm2307, Xalpharis, Xdenizen, Xerstau, Xezbeth, Xfactor2000, Xiner, Xoloz, Xyzzyva, YSSYguy, YUL89YYZ, Yah, Trick, Yah, Yahel Guhan, Yamamoto Ichiro, Yanksox, Yartch, Yasis, Yassou Jayne, Yath, Yattum, Yeaboiitskurt, YellowMonkey, Ymblanter, Yonas29, Yoooooooooork, Yopie, Yorkshirian, Youmps, Youramelon, Yozer1, Yt95, Yuckfoo, Yulia Romero, Yurko57, Yusha0450, ZRetro, ZZeta, Zaf, Zaharous, Zakiya alaamri82, Zandweb, Zaphrost, Zarniwoot, Zealander, Zeanrivera, Zedshort, Zeliboba7, Zenit4ever, Zepheriah, ZincOrbie, Zlerman, Zloyvolsheb, Zmaher, Znayka, Zoe, Zoeneo480144, Zoloto-Russland, Zombiecapper, Zondor, Zscout370, Zserghei, Zsinj, Zurkhardo, Zvar, Zz pot, Zzuuzz, Zzyzx11, Zé da Silva, Мистер ОКО, РКП, Скампецкий, Эрманарих, דוד, مساب, 3291 anonymous edits

Evenk_Autonomous_Okrug *Source*: http://en.wikipedia.org/w/index.php?title=Evenk_Autonomous_Okrug *Contributors*: Abstrakt, Aeolopluton, American Imperialist, Anandks007, Apoivre, Barend, Bjankuloski06en, Chipppy, Conscious, Darwinek, Ed Cormany, Ezhiki, Grutness, Gzornenplatz, Hottentot, IZAK, Japanese Searobin, Maximaximax, Metafactory, Morwen, Mysid,

Nickshanks, Nightstallion, Nikai, Otvis, Peyre, PhnomPencil, Plop, Rich Farmbrough, Roberta F., SM, Santiperez, Stuffisthings, Template namespace initialisation script, TheWhiteRussian, Volkovtravel, Yalens, Yannzgob, 14 anonymous edits

Federal_subjects_of_Russia *Source*: http://en.wikipedia.org/w/index.php?title=Federal_subjects_of_Russia *Contributors*: A2Kafir, ASDFGH, Abdull, Abhijitsathe, Ahmad87, Aleksandr Yazov, AlexWelens, Altes, American Imperialist, Argo Navis, Aris Katsaris, Ask123, Augur, BaF, Barend, Bart l, Betterusername, Bjankuloski06en, Bobo192, Bonzon, Brianski, Bury Medved, Carlossuarez46, Chanheigeorge, Cheezerman, ChemTerm, ClaretAsh, Colchicum, CommonsDelinker, Contributor777, Crusoe8181, Cwmacdougall, DabMachine, Dale Arnett, Darz Mol, DimaY2K, Dragquennom, Dricherby, EamonnPKeane, Eitniear, Electionworld, Emesee, Ezhiki, Feedmecereal, Folks at 137, Fotisgrek, Fremantleboy, Frysun, Glossologist, Gnomz007, Golbez, Green Giant, Greyhood, Grillo, Groovenstein, Hadrian89, Hellerick, HennessyC, Hmains, Hodja Nasreddin, HorsePunchKid, Hunnjazal, Hvn0413, IZAK, Ianare, Iridescent, Jao, Jebus989, John K, Johnluisocasio, Joy, Jørgen, KNewman, Kanags, Khajidha, Killing Vector, Kwamikagami, La goutte de pluie, Lady Pablo, Lockesdonkey, Lvivske, Masahiro Naoi, Mateom28, Mathew5000, Miaow Miaow, Miyokan, Morwen, Moscvitch, Mysteriou, Nellis, Nightstallion, Nika 243, Nixer, NuclearVacuum, Ochkarik, Otets, Paploo, Paul August, Paul Barlow, Pedro Aguiar, Piano non troppo, Pianoplonkers, Pms196, Podzemnik, Psychonaut, Qwertyus, R9tgokunks, Reywas92, Rholton, Ro4444, Roke, Rune X2, Sardanaphalus, Scs, SeNeKa, Shonyh, SiBr4, Sikon, Steakof1000names, TarzanASG, Tavork, Teemu Leisti, The President of Cool, Thefilmsmsm, Tobias Conradi, Treisijs, Uannis, Vanjagenije, Vasilcho, Voldemort175, Wassermann, Wavelength, Woohookitty, Worobiew, Wtmitchell, АбуУбайда, احمد.غامدي24, 118 anonymous edits

Administrative_centre *Source*: http://en.wikipedia.org/w/index.php?title=Administrative_centre *Contributors*: Aleksandr Grigoryev, Applefilling, Bhny, Breadandcheese, Charvex, Cncs wikipedia, DDima, Discospinster, Ezhiki, Hectorthebat, Hgb1217, Intangible, JeremyA, Jza84, MRSC, Page Up, Pilim, Pol098, Retired username, Thorsen, Tomasino96, Untifler, Whoatherekiddo, 3 anonymous edits

Types_of_inhabited_localities_in_Russia *Source*: http://en.wikipedia.org/w/index.php?title=Types_of_inhabited_localities_in_Russia *Contributors*: Abtract, Achp ru, Aleksandr Grigoryev, Altenmann, Arch dude, BD2412, Barbatus, Benjamin Trovato, Bogdan Nagachop, Dankarl, David Kernow, ENeville, Ezhiki, Gavrant, Hardscarf, Hmains, Hvn0413, Imie10, Introvert, Joy, Klimenok, Koryakov Yuri, LilHelpa, Morning Sunshine, Pro gamer 9191, Robofish, Shcha, Superzohar, The Transhumanist, TheNewPhobia, Vegaswikian, WALKER russia, Woohookitty, 12 anonymous edits

Taymyrsky_Dolgano-Nenetsky_District *Source*: http://en.wikipedia.org/w/index.php?title=Taymyrsky_Dolgano-Nenetsky_District *Contributors*: Backspace, Ezhiki, Khanson, Otvis, Vmenkov, 1 anonymous edits

Severo-Yeniseysky_District *Source*: http://en.wikipedia.org/w/index.php?title=Severo-Yeniseysky_District *Contributors*: Ezhiki, GoodDay, Khanson

Turukhansky_District *Source*: http://en.wikipedia.org/w/index.php?title=Turukhansky_District *Contributors*: Backspace, Ezhiki, Khanson, Rich Farmbrough, Vmenkov, 1 anonymous edits

Russian_Census_(2010) *Source*: http://en.wikipedia.org/w/index.php?title=Russian_Census_%282010%29 *Contributors*: Ardsarea, BD2412, CommonsDelinker, DonaldDuck, EdJohnston, El Snubbe, Ezhiki, Greyhood, Jedibob5, Jiujitsuguy, LokiiT, Lvivske, Maxim11maxim, Schekinov Alexey Victorovich, Serjio-pt, Shatter Resistance, Spb1605, Toccata quarta, Викимонетчик, 3 anonymous edits

Image Sources, Licenses and Contributors

File:KrasnoyarskKray-Flag.svg *Source*: http://en.wikipedia.org/w/index.php?title=File:KrasnoyarskKray-Flag.svg *License*: unknown *Contributors*: Hellerick

File:Coat of arms of Krasnoyarsk Krai.svg *Source*: http://en.wikipedia.org/w/index.php?title=File:Coat_of_arms_of_Krasnoyarsk_Krai.svg *License*: unknown *Contributors*: Hellerick

File:Krasnoyarsk in Russia.svg *Source*: http://en.wikipedia.org/w/index.php?title=File:Krasnoyarsk_in_Russia.svg *License*: unknown *Contributors*: TUBS

File:Krasnoyarsk Mira 110 krai admin.jpg *Source*: http://en.wikipedia.org/w/index.php?title=File:Krasnoyarsk_Mira_110_krai_admin.jpg *License*: unknown *Contributors*: Obakeneko, Siberiano, Кузнецов

File:Map of the Krasnoyarsk Krai (upright).png *Source*: http://en.wikipedia.org/w/index.php?title=File:Map_of_the_Krasnoyarsk_Krai_(upright).png *License*: unknown *Contributors*: Maximilian Dörrbecker (Chumwa)

File:Flag of Russia.svg *Source*: http://en.wikipedia.org/w/index.php?title=File:Flag_of_Russia.svg *License*: unknown *Contributors*: Anomie, Zscout370

File:Coat of Arms of the Russian Federation.svg *Source*: http://en.wikipedia.org/w/index.php?title=File:Coat_of_Arms_of_the_Russian_Federation.svg *License*: unknown *Contributors*: Федеральный конституционный закон «О Государственном гербе Российской Федерации»

File:Russian Federation (orthographic projection).svg *Source*: http://en.wikipedia.org/w/index.php?title=File:Russian_Federation_(orthographic_projection).svg *License*: unknown *Contributors*: User:Ssolbergj

File:Increase2.svg *Source*: http://en.wikipedia.org/w/index.php?title=File:Increase2.svg *License*: unknown *Contributors*: Sarang

file:speakerlink-new.svg *Source*: http://en.wikipedia.org/w/index.php?title=File:Speakerlink-new.svg *License*: unknown *Contributors*: User:Kelvinsong

File:Speaker Icon.svg *Source*: http://en.wikipedia.org/w/index.php?title=File:Speaker_Icon.svg *License*: unknown *Contributors*: Blast, G.Hagedorn, Mobius, Tehdog, 2 anonymous edits

File:IE expansion.png *Source*: http://en.wikipedia.org/w/index.php?title=File:IE_expansion.png *License*: unknown *Contributors*: Dbachmann, Esteban.barahona, Man vyi, Mdd, Mvangeest, Ricordisamoa, 1 anonymous edits

File:Kievan Rus en.jpg *Source*: http://en.wikipedia.org/w/index.php?title=File:Kievan_Rus_en.jpg *License*: unknown *Contributors*: Captain Blood, Kintetsubuffalo, Ras67, Rottweiler, Silar

File:Lebedev baptism.jpg *Source*: http://en.wikipedia.org/w/index.php?title=File:Lebedev_baptism.jpg *License*: unknown *Contributors*: Bogomolov.PL, Butko, Joseolgon, Mladifilozof, Randbewohner, Vizu, 1 anonymous edits

File:Lissner TroiceSergievaLavr.jpg *Source*: http://en.wikipedia.org/w/index.php?title=File:Lissner_TroiceSergievaLavr.jpg *License*: unknown *Contributors*: Alex Bakharev, Art-top, Dmitry Rozhkov, MishaPan, Pianist, Shakko, 1 anonymous edits

File:Ivan the Terrible (cropped).JPG *Source*: http://en.wikipedia.org/w/index.php?title=File:Ivan_the_Terrible_(cropped).JPG *License*: unknown *Contributors*: Bukk, Butko, Kürschner, Mrlopez2681, Shakko

File:Minin&Pogjarsky 2.jpg *Source*: http://en.wikipedia.org/w/index.php?title=File:Minin&Pogjarsky_2.jpg *License*: unknown *Contributors*: EugeneZelenko, Julmin, Kneiphof, Okman, 1 anonymous edits

File:Peter der-Grosse 1838.jpg *Source*: http://en.wikipedia.org/w/index.php?title=File:Peter_der-Grosse_1838.jpg *License*: unknown *Contributors*: Anathema, Butko, Darwinius, Ecummenic, Kilom691, Mathiasrex, Mattes, Miyokan, Polarlys, Sammyday, Shakko, Sparkit, TheoClarke, Vincent Steenberg, Zoloto-Russland

File:Russian Empire (orthographic projection).svg *Source*: http://en.wikipedia.org/w/index.php?title=File:Russian_Empire_(orthographic_projection).svg *License*: unknown *Contributors*: User:Shadowxfox

File:Lenintribune.jpg *Source*: http://en.wikipedia.org/w/index.php?title=File:Lenintribune.jpg *License*: unknown *Contributors*: Original uploader was Black31 at fr.wikipedia

File:Russia-2000-stamp-Tatlin Tower and Worker and Kolkhoz Woman by Vera Mukhina.jpg *Source*: http://en.wikipedia.org/w/index.php?title=File:Russia-2000-stamp-Tatlin_Tower_and_Worker_and_Kolkhoz_Woman_by_Vera_Mukhina.jpg *License*: unknown *Contributors*: Butko, Editor at Large, Gryffindor, Man vyi, Michael Romanov, Philafrenzy, Shakko, Zimin.V.G.

File:Red Army greeted in Bucharest.jpg *Source*: http://en.wikipedia.org/w/index.php?title=File:Red_Army_greeted_in_Bucharest.jpg *License*: unknown *Contributors*: Bogdan, CaptainFugu, Dahn, Jmabel, John commons, Teofilo, 2 anonymous edits

File:Convert ru kosmos077.jpg *Source*: http://en.wikipedia.org/w/index.php?title=File:Convert_ru_kosmos077.jpg *License*: unknown *Contributors*: USSR government

File:Sputnik asm.jpg *Source*: http://en.wikipedia.org/w/index.php?title=File:Sputnik_asm.jpg *License*: unknown *Contributors*: User:Jappalang

File:Moscow, City May 2010 03.JPG *Source*: http://en.wikipedia.org/w/index.php?title=File:Moscow,_City_May_2010_03.JPG *License*: unknown *Contributors*: User:NVO

File:Moscow Kremlin from Kamenny bridge.jpg *Source*: http://en.wikipedia.org/w/index.php?title=File:Moscow_Kremlin_from_Kamenny_bridge.jpg *License*: unknown *Contributors*: User:Julmin, User:Surendil

File:BRIC leaders in 2008.jpg *Source*: http://en.wikipedia.org/w/index.php?title=File:BRIC_leaders_in_2008.jpg *License*: unknown *Contributors*: Presidential Press and Information Office

File:Countries-SCO.png *Source*: http://en.wikipedia.org/w/index.php?title=File:Countries-SCO.png *License*: unknown *Contributors*: User:Sejogo

File:2010 Moscow Victory Day Parade-5.jpeg *Source*: http://en.wikipedia.org/w/index.php?title=File:2010_Moscow_Victory_Day_Parade-5.jpeg *License*: unknown *Contributors*: Presidential Press and Information Office

File:Russian Regions-EN.svg *Source*: http://en.wikipedia.org/w/index.php?title=File:Russian_Regions-EN.svg *License*: unknown *Contributors*: User:Ezhiki, User:Plavius, User:Tavork

File:Russland Relief.png *Source*: http://en.wikipedia.org/w/index.php?title=File:Russland_Relief.png *License*: unknown *Contributors*: User:Lencer

File:Elbrus North 195.jpg *Source*: http://en.wikipedia.org/w/index.php?title=File:Elbrus_North_195.jpg *License*: unknown *Contributors*: Dmitrij96, Herzi Pinki, Iltever, Kykovner, 1 anonymous edits

File:Осетр река.jpg *Source*: http://en.wikipedia.org/w/index.php?title=File:Осетр_река.jpg *License*: unknown *Contributors*: Hugo.arg, Kolossus, Lvova

File:Archangelsk taiga.JPG *Source*: http://en.wikipedia.org/w/index.php?title=File:Archangelsk_taiga.JPG *License*: unknown *Contributors*: User:Xxxl

File:Medved mzoo.jpg *Source*: http://en.wikipedia.org/w/index.php?title=File:Medved_mzoo.jpg *License*: unknown *Contributors*: User:Simm

File:WTC-Moscow.jpg *Source*: http://en.wikipedia.org/w/index.php?title=File:WTC-Moscow.jpg *License*: unknown *Contributors*: User:Wecameasromans

File:Russian economy since fall of Soviet Union.PNG *Source*: http://en.wikipedia.org/w/index.php?title=File:Russian_economy_since_fall_of_Soviet_Union.PNG *License*: unknown *Contributors*: User:LokiiT

File:Rozh.jpg *Source*: http://en.wikipedia.org/w/index.php?title=File:Rozh.jpg *License*: unknown *Contributors*: Afanasovich, Ain92, Bdk, Boo-Boo Baroo, Ghirlandajo, Ixtzib, Julia W, Le.Loup.Gris, Neukoln, Sfrandzi, Shakko, Vhorvat

File:RF NG pipestoEU.gif *Source*: http://en.wikipedia.org/w/index.php?title=File:RF_NG_pipestoEU.gif *License*: unknown *Contributors*: DOE

File:TransSiberianRailwayAtKm9288.jpg *Source*: http://en.wikipedia.org/w/index.php?title=File:TransSiberianRailwayAtKm9288.jpg *License*: unknown *Contributors*: Bukk, Bulwersator, Demidow, Ingolfson, JuTa, Krinkle, Mschlindwein, Ymblanter, 1 anonymous edits

File:NSF picture of Yamal.gif *Source*: http://en.wikipedia.org/w/index.php?title=File:NSF_picture_of_Yamal.gif *License*: unknown *Contributors*: High Contrast, Kneiphof, Maximaximax, Stunteltje, Tungsten, 1 anonymous edits

File:Lomonosovportrait.jpg *Source*: http://en.wikipedia.org/w/index.php?title=File:Lomonosovportrait.jpg *License*: unknown *Contributors*: Anne97432, Ecummenic, Hsarrazin, Jflomonosoff, Papa November, Shakko

File:Soyuz TMA-2 launch.jpg *Source*: http://en.wikipedia.org/w/index.php?title=File:Soyuz_TMA-2_launch.jpg *License*: unknown *Contributors*: NASA/Scott Andrews

File:Sukhoi T-50 Beltyukov.jpg *Source*: http://en.wikipedia.org/w/index.php?title=File:Sukhoi_T-50_Beltyukov.jpg *License*: unknown *Contributors*: Alex Beltyukov

File:Yo-conc3.JPG *Source*: http://en.wikipedia.org/w/index.php?title=File:Yo-conc3.JPG *License*: unknown *Contributors*: User:Brattarb

File:Mir on 12 June 1998edit1.jpg *Source*: http://en.wikipedia.org/w/index.php?title=File:Mir_on_12_June_1998edit1.jpg *License*: unknown *Contributors*: User:Thegreenj

Image:Percentage of Russians by region.svg *Source*: http://en.wikipedia.org/w/index.php?title=File:Percentage_of_Russians_by_region.svg *License*: unknown *Contributors*: User:Любослов Езыкин

Image:Russia natural population growth rates 2012.PNG *Source*: http://en.wikipedia.org/w/index.php?title=File:Russia_natural_population_growth_rates_2012.PNG *License*: unknown *Contributors*: User:Anvyder

File:RussianLanguageMap.png *Source*: http://en.wikipedia.org/w/index.php?title=File:RussianLanguageMap.png *License*: unknown *Contributors*: BartekChom, Chinneeb, Damian Radu, David Kernow, DenghiùComm, JuTa, Kuban kazak, Maphobbyist, NuclearVacuum, Roke, Small Bug, Spesh531, Tat1642, Wester, Winterheart, 22 anonymous edits

File:Kazan church edit1.jpg *Source*: http://en.wikipedia.org/w/index.php?title=File:Kazan_church_edit1.jpg *License*: unknown *Contributors*: Maarten

File:Terapevt Mudrov train.JPG *Source*: http://en.wikipedia.org/w/index.php?title=File:Terapevt_Mudrov_train.JPG *License*: unknown *Contributors*: User:Glucke

File:Школа 1118 (Москва).jpg *Source*: http://en.wikipedia.org/w/index.php?title=File:Школа_1118_(Москва).jpg *License*: unknown *Contributors*: vow

File:Kustodiev Merchants Wife.jpg *Source*: http://en.wikipedia.org/w/index.php?title=File:Kustodiev_Merchants_Wife.jpg *License*: unknown *Contributors*: Alex Bakharev, AndreasPraefcke, Bukk, Jonkerz, Jtir, Kaganer, Mattbuck, Mattes, Moogsi, Pierpao, Pitke, Shakko, SunOfErat, Talmoryair, Thorvaldsson, Vincent Steenberg, Wst

File:RedSquare SaintBasile (pixinn.net).jpg *Source*: http://en.wikipedia.org/w/index.php?title=File:RedSquare_SaintBasile_(pixinn.net).jpg *License*: unknown *Contributors*: User:XtoF

File:Andrej Rublëv 001.jpg *Source*: http://en.wikipedia.org/w/index.php?title=File:Andrej_Rublëv_001.jpg *License*: unknown *Contributors*: AndreasPraefcke, AnonMoos, Aroche, Goldfritha, Laurom, Mattes, Shakko, Tsca, Wst

File:Snowdance.jpg *Source*: http://en.wikipedia.org/w/index.php?title=File:Snowdance.jpg *License*: unknown *Contributors*: Rick Dikeman

File:L.N.Tolstoy Prokudin-Gorsky.jpg *Source*: http://en.wikipedia.org/w/index.php?title=File:L.N.Tolstoy_Prokudin-Gorsky.jpg *License*: unknown *Contributors*: Avenger911, Berillium, Ctac, Damouns, Ecummenic, GreatOrangePumpkin, Henry Merrivale, J.M.Domingo, Methem, MrPanyGoff, Oldcolor, Pianist, RedAndr, Reino Helismaa, Sasha Krotov, Tommy Kronkvist, Wolfmann, Yann, 3 anonymous edits

File:Medvedev - Russia Today 3.jpg *Source*: http://en.wikipedia.org/w/index.php?title=File:Medvedev_-_Russia_Today_3.jpg *License*: unknown *Contributors*: Presidential Press and Information Office

File:Shukhov tower shabolovka moscow 02.jpg *Source*: http://en.wikipedia.org/w/index.php?title=File:Shukhov_tower_shabolovka_moscow_02.jpg *License*: unknown *Contributors*: Arssenev, Basilicofresco, Donskoy, Lite, NVO, Rlevse, Roomba, Snowgrove, TomAlt, Vonvon, 5 anonymous edits

File:RussianBear.JPG *Source*: http://en.wikipedia.org/w/index.php?title=File:RussianBear.JPG *License*: unknown *Contributors*: Roman Kovrigin

File:Dmitry Medvedev 20 May 2008-2.jpg *Source*: http://en.wikipedia.org/w/index.php?title=File:Dmitry_Medvedev_20_May_2008-2.jpg *License*: unknown *Contributors*: Presidential Press and Information Office

File:Victory Day Parade 2005-18.jpg *Source*: http://en.wikipedia.org/w/index.php?title=File:Victory_Day_Parade_2005-18.jpg *License*: unknown *Contributors*: ITAR-TASS

File:Alyye parusa.jpg *Source*: http://en.wikipedia.org/w/index.php?title=File:Alyye_parusa.jpg *License*: unknown *Contributors*: Marina Lystseva

File:PeterhofGrandCascade.JPG *Source*: http://en.wikipedia.org/w/index.php?title=File:PeterhofGrandCascade.JPG *License*: unknown *Contributors*: Cottbus, Kaganer, Ronaldino

File:Russian-Matroshka2.jpg *Source*: http://en.wikipedia.org/w/index.php?title=File:Russian-Matroshka2.jpg *License*: unknown *Contributors*: Fanghong, Ingolfson, Mariluna

File:Gnome-globe.svg *Source*: http://en.wikipedia.org/w/index.php?title=File:Gnome-globe.svg *License*: unknown *Contributors*: David Vignoni

File:RussiaEvenkia2005.png *Source*: http://en.wikipedia.org/w/index.php?title=File:RussiaEvenkia2005.png *License*: unknown *Contributors*: Bencmq, Ezhiki, Imz, Maximaximax, PANONIAN

File:Flag of Evenkia.svg *Source*: http://en.wikipedia.org/w/index.php?title=File:Flag_of_Evenkia.svg *License*: unknown *Contributors*: Mysid

File:Federal subjects of Russia (by number).png *Source*: http://en.wikipedia.org/w/index.php?title=File:Federal_subjects_of_Russia_(by_number).png *License*: unknown *Contributors*: Fremantleboy, Legnaw, Maximaximax, Nightstallion, Red cz, Romanm, Stannered, 1 anonymous edits

File:Flag of Adygea.svg *Source*: http://en.wikipedia.org/w/index.php?title=File:Flag_of_Adygea.svg *License*: unknown *Contributors*: User:Conscious

File:Adygeya - Coat of Arms.png *Source*: http://en.wikipedia.org/w/index.php?title=File:Adygeya_-_Coat_of_Arms.png *License*: unknown *Contributors*: Blast, Hellerick, Kilom691, Magul, Mattes, Yuriy75, Óðinn, Закубанец

File:Flag of Bashkortostan.svg *Source*: http://en.wikipedia.org/w/index.php?title=File:Flag_of_Bashkortostan.svg *License*: unknown *Contributors*: User:Conscious

File:Coat of Arms of Bashkortostan.svg *Source*: http://en.wikipedia.org/w/index.php?title=File:Coat_of_Arms_of_Bashkortostan.svg *License*: unknown *Contributors*: Fazletdin Islakhov

File:Flag of Buryatia.svg *Source*: http://en.wikipedia.org/w/index.php?title=File:Flag_of_Buryatia.svg *License*: unknown *Contributors*: A Werewolf, Erlenmeyer, Fred the Oyster, Maks Stirlitz, Mattes, NVO, Orange Tuesday, Scriber en, Urmas

File:Coat of Arms of Buryatiya.svg *Source*: http://en.wikipedia.org/w/index.php?title=File:Coat_of_Arms_of_Buryatiya.svg *License*: unknown *Contributors*: Art-top, Homo lupus, NVO, Óðinn

File:Flag of Altai Republic.svg *Source*: http://en.wikipedia.org/w/index.php?title=File:Flag_of_Altai_Republic.svg *License*: unknown *Contributors*: Gauravjuvekar, Kookaburra, Maks Stirlitz, Mattes, Obakeneko, Permjak, Prev, Scriber en, Urmas, 1 anonymous edits

File:Coat of Arms of Altai Republic.svg *Source*: http://en.wikipedia.org/w/index.php?title=File:Coat_of_Arms_of_Altai_Republic.svg *License*: unknown *Contributors*: unknown (vectorized Perhelion)

File:Flag of Dagestan.svg *Source*: http://en.wikipedia.org/w/index.php?title=File:Flag_of_Dagestan.svg *License*: unknown *Contributors*: User:Stianbh

File:Coat of Arms of Dagestan.svg *Source*: http://en.wikipedia.org/w/index.php?title=File:Coat_of_Arms_of_Dagestan.svg *License*: unknown *Contributors*: Art-top, Cycn, Darwinius, Óðinn

File:Flag of Ingushetia.svg *Source*: http://en.wikipedia.org/w/index.php?title=File:Flag_of_Ingushetia.svg *License*: unknown *Contributors*: Alex Spade, AnonMoos, Artem Karimov, Chtak Yuno, Fred the Oyster, Gauravjuvekar, Incnis Mrsi, Maks Stirlitz, Mattes, McZusatz, Mysid, Permjak, Ratatosk, Scriber en, Stianbh, Thuresson, Waldir, Yuriy75, 1 anonymous edits

File:Coat of Arms of Ingushetia.svg *Source*: http://en.wikipedia.org/w/index.php?title=File:Coat_of_Arms_of_Ingushetia.svg *License*: unknown *Contributors*: Ruslan Eldiyev

File:Flag of Kabardino-Balkaria.svg *Source*: http://en.wikipedia.org/w/index.php?title=File:Flag_of_Kabardino-Balkaria.svg *License*: unknown *Contributors*: User:Conscious

File:Coat of Arms of Kabardino-Balkaria.svg *Source*: http://en.wikipedia.org/w/index.php?title=File:Coat_of_Arms_of_Kabardino-Balkaria.svg *License*: unknown *Contributors*: Yakub Akkizov, Gherman Pashtov

File:Flag of Kalmykia.svg *Source*: http://en.wikipedia.org/w/index.php?title=File:Flag_of_Kalmykia.svg *License*: unknown *Contributors*: User:Conscious

File:Coat of Arms of Kalmykia.svg *Source*: http://en.wikipedia.org/w/index.php?title=File:Coat_of_Arms_of_Kalmykia.svg *License*: unknown *Contributors*: Bata Erdniyev

File:Flag of Karachay-Cherkessia.svg *Source*: http://en.wikipedia.org/w/index.php?title=File:Flag_of_Karachay-Cherkessia.svg *License*: unknown *Contributors*: User:Conscious

File:Coat of Arms of Karachay-Cherkessia.svg *Source*: http://en.wikipedia.org/w/index.php?title=File:Coat_of_Arms_of_Karachay-Cherkessia.svg *License*: unknown *Contributors*: Umar Mizhev

File:Flag of Karelia.svg *Source*: http://en.wikipedia.org/w/index.php?title=File:Flag_of_Karelia.svg *License*: unknown *Contributors*: User:Fibonacci

File:Coat of Arms of Republic of Karelia.svg *Source*: http://en.wikipedia.org/w/index.php?title=File:Coat_of_Arms_of_Republic_of_Karelia.svg *License*: unknown *Contributors*: Yuri Nivin

File:Flag of Komi.svg *Source*: http://en.wikipedia.org/w/index.php?title=File:Flag_of_Komi.svg *License*: unknown *Contributors*: В. Я. Сердитов

File:Coat of Arms of the Komi Republic.svg *Source*: http://en.wikipedia.org/w/index.php?title=File:Coat_of_Arms_of_the_Komi_Republic.svg *License*: unknown *Contributors*: A. Neverov

File:Flag of Mari El.svg *Source*: http://en.wikipedia.org/w/index.php?title=File:Flag_of_Mari_El.svg *License*: unknown *Contributors*: User:Urmas

File:Coat of Arms of Mari El.svg *Source*: http://en.wikipedia.org/w/index.php?title=File:Coat_of_Arms_of_Mari_El.svg *License*: unknown *Contributors*: Art-top, Gajmar, Hellerick, Zscout370, Óðinn

File:Flag of Mordovia.svg *Source*: http://en.wikipedia.org/w/index.php?title=File:Flag_of_Mordovia.svg *License*: unknown *Contributors*: User:Fibonacci, User:Fibonacci

File:Coat of Arms of Mordovia.svg *Source*: http://en.wikipedia.org/w/index.php?title=File:Coat_of_Arms_of_Mordovia.svg *License*: unknown *Contributors*: Nikolai Chikrinyov

File:Flag of Sakha.svg *Source*: http://en.wikipedia.org/w/index.php?title=File:Flag_of_Sakha.svg *License*: unknown *Contributors*: Gauravjuvekar, Johnny Rotten, Mattes, NuclearVacuum, Urmas, 1 anonymous edits

File:Coat of Arms of Sakha (Yakutia).svg *Source*: http://en.wikipedia.org/w/index.php?title=File:Coat_of_Arms_of_Sakha_(Yakutia).svg *License*: unknown *Contributors*: User:Azgar

File:Flag of North Ossetia.svg *Source*: http://en.wikipedia.org/w/index.php?title=File:Flag_of_North_Ossetia.svg *License*: unknown *Contributors*: Anime Addict AA, AnonMoos, Artem Karimov, Erlenmeyer, Fry1989, Homo lupus, Jedan02, Kookaburra, Masseman, Mattes, Nachcommonsverschieber, Orange-kun, Pumbaa80, Ricordisamoa, Urmas, Yuriy75, Егор Осин

File:Coat of Arms of North Ossetia-Alania.png *Source*: http://en.wikipedia.org/w/index.php?title=File:Coat_of_Arms_of_North_Ossetia-Alania.png *License*: unknown *Contributors*: Denis Syabukov

File:Flag of Tatarstan.svg *Source*: http://en.wikipedia.org/w/index.php?title=File:Flag_of_Tatarstan.svg *License*: unknown *Contributors*: User:Fibonacci

File:Coat of Arms of Tatarstan.svg *Source*: http://en.wikipedia.org/w/index.php?title=File:Coat_of_Arms_of_Tatarstan.svg *License*: unknown *Contributors*: I. Zagitov, N. Khanzafarov, E. Saginov, R. Fakhrutdinov

File:Flag of Tuva.svg *Source*: http://en.wikipedia.org/w/index.php?title=File:Flag_of_Tuva.svg *License*: unknown *Contributors*: Mysid

File:Coat of arms of Tuva.svg *Source*: http://en.wikipedia.org/w/index.php?title=File:Coat_of_arms_of_Tuva.svg *License*: unknown *Contributors*: Mysid

File:Flag of Udmurtia.svg *Source*: http://en.wikipedia.org/w/index.php?title=File:Flag_of_Udmurtia.svg *License*: unknown *Contributors*: User:Fibonacci

File:Coat of arms of Udmurtia.svg *Source*: http://en.wikipedia.org/w/index.php?title=File:Coat_of_arms_of_Udmurtia.svg *License*: unknown *Contributors*: [Ю.Н.Лобанов]]

File:Flag of Khakassia.svg *Source*: http://en.wikipedia.org/w/index.php?title=File:Flag_of_Khakassia.svg *License*: unknown *Contributors*: User:Fibonacci

File:Coat of arms of Khakassia.svg *Source*: http://en.wikipedia.org/w/index.php?title=File:Coat_of_arms_of_Khakassia.svg *License*: unknown *Contributors*: Aleksandr Kotozhekov

File:Coat of arms of Chechnya.svg *Source*: http://en.wikipedia.org/w/index.php?title=File:Coat_of_arms_of_Chechnya.svg *License*: unknown *Contributors*: User:Anuskafm, User:ChristianBier

File:Flag of Chuvashia.svg *Source*: http://en.wikipedia.org/w/index.php?title=File:Flag_of_Chuvashia.svg *License*: unknown *Contributors*: User:....

File:Coat of Arms of Chuvashia.svg *Source*: http://en.wikipedia.org/w/index.php?title=File:Coat_of_Arms_of_Chuvashia.svg *License*: unknown *Contributors*: E. Yuriev

File:Flag of Altai Krai.svg *Source*: http://en.wikipedia.org/w/index.php?title=File:Flag_of_Altai_Krai.svg *License*: unknown *Contributors*: User:Дмитрий-5-Аверин

File:Coat of Arms of Altai Krai.svg *Source*: http://en.wikipedia.org/w/index.php?title=File:Coat_of_Arms_of_Altai_Krai.svg *License*: unknown *Contributors*: ANKAWÜ, Ahonc, Art-top, Óðinn

File:Flag_of_Zabaykalsky_Krai.svg *Source*: http://en.wikipedia.org/w/index.php?title=File:Flag_of_Zabaykalsky_Krai.svg *License*: unknown *Contributors*: User:Walden69

File:Coat of arms of Zabaykalsky Krai.svg *Source*: http://en.wikipedia.org/w/index.php?title=File:Coat_of_arms_of_Zabaykalsky_Krai.svg *License*: unknown *Contributors*: De728631, Popadius

File:Flag of Kamchatka Krai.svg *Source*: http://en.wikipedia.org/w/index.php?title=File:Flag_of_Kamchatka_Krai.svg *License*: unknown *Contributors*: User:Permjak

File:Coat of Arms of Kamchatka Krai.svg *Source*: http://en.wikipedia.org/w/index.php?title=File:Coat_of_Arms_of_Kamchatka_Krai.svg *License*: unknown *Contributors*: ANKAWÜ, Art-top, Хинт

File:Flag of Krasnodar Krai.png *Source*: http://en.wikipedia.org/w/index.php?title=File:Flag_of_Krasnodar_Krai.png *License*: unknown *Contributors*: Aivazovsky, Hellerick, Scriber en, Strannik27, Yuriy75

File:Coat of Arms of Krasnodar kray.png *Source*: http://en.wikipedia.org/w/index.php?title=File:Coat_of_Arms_of_Krasnodar_kray.png *License*: unknown *Contributors*: Hellerick, Magul, Maximaximax, Panther, Yuriy75, Óðinn

File:Flag of Perm Krai.svg *Source*: http://en.wikipedia.org/w/index.php?title=File:Flag_of_Perm_Krai.svg *License*: unknown *Contributors*: User:const_st

File:Coat of Arms of Perm Krai.svg *Source*: http://en.wikipedia.org/w/index.php?title=File:Coat_of_Arms_of_Perm_Krai.svg *License*: unknown *Contributors*: Margarita Tarasova; based on historical coat of arms, adopted in 1783

File:Flag of Primorsky Krai.svg *Source*: http://en.wikipedia.org/w/index.php?title=File:Flag_of_Primorsky_Krai.svg *License*: unknown *Contributors*: Conscious, Maks Stirlitz, Mattes, NVO

File:Coat of arms of Primorsky Krai.svg *Source*: http://en.wikipedia.org/w/index.php?title=File:Coat_of_arms_of_Primorsky_Krai.svg *License*: unknown *Contributors*: Arachn0, Conscious, Maks Stirlitz, NVO

File:Flag of Stavropol Krai.png *Source*: http://en.wikipedia.org/w/index.php?title=File:Flag_of_Stavropol_Krai.png *License*: unknown *Contributors*: User:Conscious

File:Coat of Arms of Stavropol kray.png *Source*: http://en.wikipedia.org/w/index.php?title=File:Coat_of_Arms_of_Stavropol_kray.png *License*: unknown *Contributors*: Hellerick, Magul, Maximaximax, Panther, Yuriy75, Óðinn

File:Flag of Khabarovsk Krai.svg *Source*: http://en.wikipedia.org/w/index.php?title=File:Flag_of_Khabarovsk_Krai.svg *License*: unknown *Contributors*: User:Fibonacci

File:Coat of Arms of Khabarovsky kray (N2).png *Source*: http://en.wikipedia.org/w/index.php?title=File:Coat_of_Arms_of_Khabarovsky_kray_(N2).png *License*: unknown *Contributors*: Kobac, Maximaximax, Panther

File:Flag of Amur Oblast.svg *Source*: http://en.wikipedia.org/w/index.php?title=File:Flag_of_Amur_Oblast.svg *License*: unknown *Contributors*: User:Conscious

File:Coat of Arms of Amur oblast.png *Source*: http://en.wikipedia.org/w/index.php?title=File:Coat_of_Arms_of_Amur_oblast.png *License*: unknown *Contributors*: Magul, Panther

File:Flag of Arkhangelsk Oblast.svg *Source*: http://en.wikipedia.org/w/index.php?title=File:Flag_of_Arkhangelsk_Oblast.svg *License*: unknown *Contributors*: Закон Архангельской области от 23 сентября 2009 г. N 56-5-ОЗ "О флаге Архангельской области"

File:Coat of Arms of Arkhangelsk oblast.svg *Source*: http://en.wikipedia.org/w/index.php?title=File:Coat_of_Arms_of_Arkhangelsk_oblast.svg *License*: unknown *Contributors*: Закон Архангельской области от 15 июля 2003 г. N 178-23-ОЗ "О гербе Архангельской области"

File:Flag of Astrakhan Oblast.svg *Source*: http://en.wikipedia.org/w/index.php?title=File:Flag_of_Astrakhan_Oblast.svg *License*: unknown *Contributors*: User:Conscious

File:Coat of Arms of Astrakhan Oblast.png *Source*: http://en.wikipedia.org/w/index.php?title=File:Coat_of_Arms_of_Astrakhan_Oblast.png *License*: unknown *Contributors*: DenghiùComm, Himasaram, Kilom691, NVO, Sasha Krotov, Stianbh, Ysangkok, 1 anonymous edits

File:Flag of Belgorod Oblast.svg *Source*: http://en.wikipedia.org/w/index.php?title=File:Flag_of_Belgorod_Oblast.svg *License*: unknown *Contributors*: Hellerick, Mattes, NVO, Scriber en

File:Lob Coat of Arms of Belgorod Oblast.svg *Source*: http://en.wikipedia.org/w/index.php?title=File:Lob_Coat_of_Arms_of_Belgorod_Oblast.svg *License*: unknown *Contributors*: User:Лобачев Владимир

File:Flag of Bryansk Oblast.svg *Source*: http://en.wikipedia.org/w/index.php?title=File:Flag_of_Bryansk_Oblast.svg *License*: unknown *Contributors*: User:Cian Akril

File:Coat of Arms of Bryansk Oblast.svg *Source*: http://en.wikipedia.org/w/index.php?title=File:Coat_of_Arms_of_Bryansk_Oblast.svg *License*: unknown *Contributors*: User:Cian Akril

File:Flag of Vladimirskaya Oblast.svg *Source*: http://en.wikipedia.org/w/index.php?title=File:Flag_of_Vladimirskaya_Oblast.svg *License*: unknown *Contributors*: User:Cian Akril

File:Coat of arms of Vladimiri Oblast.svg *Source*: http://en.wikipedia.org/w/index.php?title=File:Coat_of_arms_of_Vladimiri_Oblast.svg *License*: unknown *Contributors*: User:Cian Akril

File:Flag of Volgograd Oblast.svg *Source*: http://en.wikipedia.org/w/index.php?title=File:Flag_of_Volgograd_Oblast.svg *License*: unknown *Contributors*: Cian Akril, Hellerick, Redboston, Russavia

File:Coat of Arms of Volgograd oblast.svg *Source*: http://en.wikipedia.org/w/index.php?title=File:Coat_of_Arms_of_Volgograd_oblast.svg *License*: unknown *Contributors*: User:Cian Akril

File:Flag of Vologda oblast.svg *Source*: http://en.wikipedia.org/w/index.php?title=File:Flag_of_Vologda_oblast.svg *License*: unknown *Contributors*: Cycn, Popadius, Ricordisamoa

File:Coat of arms of Vologda oblast.svg *Source*: http://en.wikipedia.org/w/index.php?title=File:Coat_of_arms_of_Vologda_oblast.svg *License*: unknown *Contributors*: Okras, Popadius, Putnik

File:Flag of Voronezh Oblast.svg *Source*: http://en.wikipedia.org/w/index.php?title=File:Flag_of_Voronezh_Oblast.svg *License*: unknown *Contributors*: Mysid after António Martins

File:Coat of Arms of Voronezh oblast (2005).png *Source*: http://en.wikipedia.org/w/index.php?title=File:Coat_of_Arms_of_Voronezh_oblast_(2005).png *License*: unknown *Contributors*: K. Mochyonov, Yu. Korzhik, K. Medvedev, R. Malanichev, O. Afanasyeva

File:Flag of Ivanovo Oblast.png *Source*: http://en.wikipedia.org/w/index.php?title=File:Flag_of_Ivanovo_Oblast.png *License*: unknown *Contributors*: User:Conscious

File:Coat of Arms of Ivanovo oblast.png *Source*: http://en.wikipedia.org/w/index.php?title=File:Coat_of_Arms_of_Ivanovo_oblast.png *License*: unknown *Contributors*: Hellerick, Magul, Maximaximax, NVO, Panther, Пакко

File:Flag of Irkutsk Oblast.svg *Source*: http://en.wikipedia.org/w/index.php?title=File:Flag_of_Irkutsk_Oblast.svg *License*: unknown *Contributors*: EJavanainen, NVO, Saibo, Scriber en

File:Coat of arms of Irkutsk Oblast.png *Source*: http://en.wikipedia.org/w/index.php?title=File:Coat_of_arms_of_Irkutsk_Oblast.png *License*: unknown *Contributors*: Gronchih, Hellerick, Magul, NVO, Stianbh

File:Flag of Kaliningrad Oblast.svg *Source*: http://en.wikipedia.org/w/index.php?title=File:Flag_of_Kaliningrad_Oblast.svg *License*: unknown *Contributors*: Original uploader was RaviC at en.wikipedia

File:Kaliningrad Oblast Coat of Arms 2006.svg *Source*: http://en.wikipedia.org/w/index.php?title=File:Kaliningrad_Oblast_Coat_of_Arms_2006.svg *License*: unknown *Contributors*: User:Bitplane

File:Flag of Kaluga Oblast.svg *Source*: http://en.wikipedia.org/w/index.php?title=File:Flag_of_Kaluga_Oblast.svg *License*: unknown *Contributors*: User:Пакко

File:Gerb kalug obl.png *Source*: http://en.wikipedia.org/w/index.php?title=File:Gerb_kalug_obl.png *License*: unknown *Contributors*: Demidow, Editor at Large, Kneiphof, Stianbh

File:Flag of Kemerovo oblast.svg *Source*: http://en.wikipedia.org/w/index.php?title=File:Flag_of_Kemerovo_oblast.svg *License*: unknown *Contributors*: Bukk, Cycn, NVO, Permjak, Пакко

File:Coat of arms of Kemerovo Oblast.svg *Source*: http://en.wikipedia.org/w/index.php?title=File:Coat_of_arms_of_Kemerovo_Oblast.svg *License*: unknown *Contributors*: User:Пакко

File:Flag of Kirov Region.svg *Source*: http://en.wikipedia.org/w/index.php?title=File:Flag_of_Kirov_Region.svg *License*: unknown *Contributors*: User:Lzhl

File:Coat of arms of Kirov Region.svg *Source*: http://en.wikipedia.org/w/index.php?title=File:Coat_of_arms_of_Kirov_Region.svg *License*: unknown *Contributors*: User:Lzhl

File:Flag of kostroma oblast.gif *Source*: http://en.wikipedia.org/w/index.php?title=File:Flag_of_kostroma_oblast.gif *License*: unknown *Contributors*: user:Hardscarf

File:Coat of arms of Kostroma oblast.gif *Source*: http://en.wikipedia.org/w/index.php?title=File:Coat_of_arms_of_Kostroma_oblast.gif *License*: unknown *Contributors*: User:ChristianBier, User:Loyna

File:Flag of Kurgan Oblast.svg *Source*: http://en.wikipedia.org/w/index.php?title=File:Flag_of_Kurgan_Oblast.svg *License*: unknown *Contributors*: User:Conscious

File:Coat of Arms of Kurgan oblast.png *Source*: http://en.wikipedia.org/w/index.php?title=File:Coat_of_Arms_of_Kurgan_oblast.png *License*: unknown *Contributors*: A.Savin, Lady Pablo, Magul, Maximaximax, Panther, Георгий Долгопский

File:Flag of Kursk Oblast.png *Source*: http://en.wikipedia.org/w/index.php?title=File:Flag_of_Kursk_Oblast.png *License*: unknown *Contributors*: Anime Addict AA, Conscious, Hellerick, Himasaram, Homo lupus, Mattes, Ricordisamoa

File:Coat of Arms of Kursk oblast.png *Source*: http://en.wikipedia.org/w/index.php?title=File:Coat_of_Arms_of_Kursk_oblast.png *License*: unknown *Contributors*: разработан на базе герба Курской губернии (1878), в основе которого лежит герб Курска известный еще по Знаменному гербовнику 1730 г.

File:Flag of Leningrad Oblast.svg *Source*: http://en.wikipedia.org/w/index.php?title=File:Flag_of_Leningrad_Oblast.svg *License*: unknown *Contributors*: User:Conscious

File:Coat of arms of Leningrad Oblast.svg *Source*: http://en.wikipedia.org/w/index.php?title=File:Coat_of_arms_of_Leningrad_Oblast.svg *License*: unknown *Contributors*: User:Conscious

File:Flag of Lipetsk Oblast.svg *Source*: http://en.wikipedia.org/w/index.php?title=File:Flag_of_Lipetsk_Oblast.svg *License*: unknown *Contributors*: Popadius, Ricordisamoa, Tnpnik

File:Coat of Arms of Lipetsk oblast.svg *Source*: http://en.wikipedia.org/w/index.php?title=File:Coat_of_Arms_of_Lipetsk_oblast.svg *License*: unknown *Contributors*: User:Stinec

File:Flag of Magadan Oblast.svg *Source*: http://en.wikipedia.org/w/index.php?title=File:Flag_of_Magadan_Oblast.svg *License*: unknown *Contributors*: Cycn, Popadius, Ricordisamoa

File:Coat of Arms of Magadan oblast.svg *Source*: http://en.wikipedia.org/w/index.php?title=File:Coat_of_Arms_of_Magadan_oblast.svg *License*: unknown *Contributors*: Popadius,

File:Flag of Moscow oblast.svg *Source*: http://en.wikipedia.org/w/index.php?title=File:Flag_of_Moscow_oblast.svg *License*: unknown *Contributors*: Artem Karimov, Arvedui89, Cycn, Homo lupus, INS Pirat, Jahobr, Jedan02, NVO, Rocket000

File:Coat of Arms of Moscow oblast.png *Source*: http://en.wikipedia.org/w/index.php?title=File:Coat_of_Arms_of_Moscow_oblast.png *License*: unknown *Contributors*: DenghiùComm, Maximaximax, NVO, Panther, Sputnik, Ysangkok, Юкатан, 1 anonymous edits

File:Flag of Murmansk Oblast.svg *Source*: http://en.wikipedia.org/w/index.php?title=File:Flag_of_Murmansk_Oblast.svg *License*: unknown *Contributors*: User:Conscious

File:Coat of Arms of Murmansk oblast (2004).png *Source*: http://en.wikipedia.org/w/index.php?title=File:Coat_of_Arms_of_Murmansk_oblast_(2004).png *License*: unknown *Contributors*: Conscious, Insider, Maximaximax, Panther, Shizhao, 2 anonymous edits

File:Flag of Nizhny Novgorod Region.svg *Source*: http://en.wikipedia.org/w/index.php?title=File:Flag_of_Nizhny_Novgorod_Region.svg *License*: unknown *Contributors*: Евгений Катышев

File:Coat of arms of Nizhny Novgorod Region.svg *Source*: http://en.wikipedia.org/w/index.php?title=File:Coat_of_arms_of_Nizhny_Novgorod_Region.svg *License*: unknown *Contributors*: Евгений Катышев

File:Flag of Novgorod Oblast.svg *Source*: http://en.wikipedia.org/w/index.php?title=File:Flag_of_Novgorod_Oblast.svg *License*: unknown *Contributors*: User:Loginname

File:Coat of arms of Novgorod Oblast.svg *Source*: http://en.wikipedia.org/w/index.php?title=File:Coat_of_arms_of_Novgorod_Oblast.svg *License*: unknown *Contributors*: User:Loginname

File:Flag of Novosibirsk oblast.svg *Source*: http://en.wikipedia.org/w/index.php?title=File:Flag_of_Novosibirsk_oblast.svg *License*: unknown *Contributors*: User:Nokka

File:Coat of arms of Novosibirsk oblast.svg *Source*: http://en.wikipedia.org/w/index.php?title=File:Coat_of_arms_of_Novosibirsk_oblast.svg *License*: unknown *Contributors*: User:Nokka

File:Flag of Omsk Oblast.svg *Source*: http://en.wikipedia.org/w/index.php?title=File:Flag_of_Omsk_Oblast.svg *License*: unknown *Contributors*: User:Conscious

File:Coat of arms of Omsk Oblast.png *Source*: http://en.wikipedia.org/w/index.php?title=File:Coat_of_arms_of_Omsk_Oblast.png *License*: unknown *Contributors*: User:Superm401, User:Óðinn

File:Flag of Orenburg Oblast.png *Source*: http://en.wikipedia.org/w/index.php?title=File:Flag_of_Orenburg_Oblast.png *License*: unknown *Contributors*: vector-images.com

File:Coat of Arms of Orenburg oblast.png *Source*: http://en.wikipedia.org/w/index.php?title=File:Coat_of_Arms_of_Orenburg_oblast.png *License*: unknown *Contributors*: Magul, Maximaximax, Panther

File:Flag of Oryol Oblast.png *Source*: http://en.wikipedia.org/w/index.php?title=File:Flag_of_Oryol_Oblast.png *License*: unknown *Contributors*: Conscious, Magul, Maks Stirlitz, Permjak

File:Coat of arms of Oryol oblast.png *Source*: http://en.wikipedia.org/w/index.php?title=File:Coat_of_arms_of_Oryol_oblast.png *License*: unknown *Contributors*: unknown

File:Flag of Penza Oblast.png *Source*: http://en.wikipedia.org/w/index.php?title=File:Flag_of_Penza_Oblast.png *License*: unknown *Contributors*: Conscious, Cycn, Himasaram, Homo lupus, Maks Stirlitz, Mattes, Pauk

File:Coat of Arms of Penza oblast.png *Source*: http://en.wikipedia.org/w/index.php?title=File:Coat_of_Arms_of_Penza_oblast.png *License*: unknown *Contributors*: Magul, Maximaximax, Panther

File:Flag placeholder.svg *Source*: http://en.wikipedia.org/w/index.php?title=File:Flag_placeholder.svg *License*: unknown *Contributors*: User:Andrwsc

File:Coat of Arms of Pskov oblast.png *Source*: http://en.wikipedia.org/w/index.php?title=File:Coat_of_Arms_of_Pskov_oblast.png *License*: unknown *Contributors*: Arthur Diebold, Juergenk59, Magul, Maximaximax, NVO, Ninrouter, Panther

File:Flag of Rostov Oblast.svg *Source*: http://en.wikipedia.org/w/index.php?title=File:Flag_of_Rostov_Oblast.svg *License*: unknown *Contributors*: User:Lokal Profil, User:Walden69

File:Rostov oblast coa.png *Source*: http://en.wikipedia.org/w/index.php?title=File:Rostov_oblast_coa.png *License*: unknown *Contributors*: Doff, Stianbh

File:Flag of Ryazan Oblast.png *Source*: http://en.wikipedia.org/w/index.php?title=File:Flag_of_Ryazan_Oblast.png *License*: unknown *Contributors*: vector-images.com

File:Coat of Arms of Ryazan oblast.png *Source*: http://en.wikipedia.org/w/index.php?title=File:Coat_of_Arms_of_Ryazan_oblast.png *License*: unknown *Contributors*: De728631, Demidow, Magul, Maximaximax, NBS, Panther, Plush-centaur

File:Flag of Samara Oblast.svg *Source*: http://en.wikipedia.org/w/index.php?title=File:Flag_of_Samara_Oblast.svg *License*: unknown *Contributors*: AnonMoos, Conscious, Cycn, Denelson83, Homo lupus, Mattes, NVO, Nightstallion, Torsch

File:Coat of Arms of Samara oblast.png *Source*: http://en.wikipedia.org/w/index.php?title=File:Coat_of_Arms_of_Samara_oblast.png *License*: unknown *Contributors*: Magul, Maximaximax, NVO, Panther

File:Flag of Saratov Oblast.png *Source*: http://en.wikipedia.org/w/index.php?title=File:Flag_of_Saratov_Oblast.png *License*: unknown *Contributors*: Bryan Derksen, Conscious, Maks Stirlitz, Mattes, NVO, Odder

File:Coat of Arms of Saratov oblast.svg *Source*: http://en.wikipedia.org/w/index.php?title=File:Coat_of_Arms_of_Saratov_oblast.svg *License*: unknown *Contributors*: User:Kaidor

File:Flag of Sakhalin Oblast.svg *Source*: http://en.wikipedia.org/w/index.php?title=File:Flag_of_Sakhalin_Oblast.svg *License*: unknown *Contributors*: User:Conscious

File:Sakhalin Oblast Coat of Arms.png *Source*: http://en.wikipedia.org/w/index.php?title=File:Sakhalin_Oblast_Coat_of_Arms.png *License*: unknown *Contributors*: not specified

File:Flag of Sverdlovsk Oblast.svg *Source*: http://en.wikipedia.org/w/index.php?title=File:Flag_of_Sverdlovsk_Oblast.svg *License*: unknown *Contributors*: User:Conscious

File:Coat of Arms of Sverdlovsk oblast.svg *Source*: http://en.wikipedia.org/w/index.php?title=File:Coat_of_Arms_of_Sverdlovsk_oblast.svg *License*: unknown *Contributors*: Cian Akril

File:Flag of Smolensk oblast.svg *Source*: http://en.wikipedia.org/w/index.php?title=File:Flag_of_Smolensk_oblast.svg *License*: unknown *Contributors*: Popadius

File:Coat of arms of Smolensk oblast.svg *Source*: http://en.wikipedia.org/w/index.php?title=File:Coat_of_arms_of_Smolensk_oblast.svg *License*: unknown *Contributors*: Popadius,

File:Flag of Tambov Oblast.svg *Source*: http://en.wikipedia.org/w/index.php?title=File:Flag_of_Tambov_Oblast.svg *License*: unknown *Contributors*: Cian Akril, Conscious, Cycn, Homo lupus

File:Coat of arms of Tambov Oblast.svg *Source*: http://en.wikipedia.org/w/index.php?title=File:Coat_of_arms_of_Tambov_Oblast.svg *License*: unknown *Contributors*: Konstantin Mochenov, Yuri Korzhik, Robert Malanichev, Galina Tunik

File:Flag of Tver Oblast.png *Source*: http://en.wikipedia.org/w/index.php?title=File:Flag_of_Tver_Oblast.png *License*: unknown *Contributors*: Anime Addict AA, Bryan Derksen, Conscious, Cycn, Maks Stirlitz, Mattes, NVO, Odder, Popadius, Ricordisamoa, Zac Allan, 1 anonymous edits

File:Coat of Arms of Tver oblast.png *Source*: http://en.wikipedia.org/w/index.php?title=File:Coat_of_Arms_of_Tver_oblast.png *License*: unknown *Contributors*: Magul, Maximaximax, NVO, Panther, Xalmeidax, 1 anonymous edits

File:TomskOblastFlag.png *Source*: http://en.wikipedia.org/w/index.php?title=File:TomskOblastFlag.png *License*: unknown *Contributors*: ChristianBier, CommonsDelinker, Darwinius, Figure19, Little Savage, Maks Stirlitz, Maximaximax, Permjak, Pitke, Scriber en, Stianbh

File:Coat of arms of Tomsk Oblast.png *Source*: http://en.wikipedia.org/w/index.php?title=File:Coat_of_arms_of_Tomsk_Oblast.png *License*: unknown *Contributors*: User:ChristianBier, User:Maximaximax

File:Flag of Tula Oblast.svg *Source*: http://en.wikipedia.org/w/index.php?title=File:Flag_of_Tula_Oblast.svg *License*: unknown *Contributors*: User:Conscious

File:Coat of Arms of Tula oblast.png *Source*: http://en.wikipedia.org/w/index.php?title=File:Coat_of_Arms_of_Tula_oblast.png *License*: unknown *Contributors*: Magul, Maximaximax, Panther, Usama

File:Flag of Tyumen Oblast.svg *Source*: http://en.wikipedia.org/w/index.php?title=File:Flag_of_Tyumen_Oblast.svg *License*: unknown *Contributors*: User:Conscious

File:Coat of arms of Tyumen Oblast.svg *Source*: http://en.wikipedia.org/w/index.php?title=File:Coat_of_arms_of_Tyumen_Oblast.svg *License*: unknown *Contributors*: Mysid

File:Flag of Ulyanovsk Oblast.png *Source*: http://en.wikipedia.org/w/index.php?title=File:Flag_of_Ulyanovsk_Oblast.png *License*: unknown *Contributors*: Conscious, Cycn, Himasaram, Homo lupus, Maks Stirlitz, Mattes, NVO, Torsch

File:Coat of Arms of Ulyanovsk Oblast.png *Source*: http://en.wikipedia.org/w/index.php?title=File:Coat_of_Arms_of_Ulyanovsk_Oblast.png *License*: unknown *Contributors*: Amdf, Magul, NVO

File:Flag of Chelyabinsk Oblast.svg *Source*: http://en.wikipedia.org/w/index.php?title=File:Flag_of_Chelyabinsk_Oblast.svg *License*: unknown *Contributors*: Mysid

File:Coat of arms of Chelyabinsk Oblast.svg *Source*: http://en.wikipedia.org/w/index.php?title=File:Coat_of_arms_of_Chelyabinsk_Oblast.svg *License*: unknown *Contributors*: Mysid

File:Flag of Yaroslavl Oblast.svg *Source*: http://en.wikipedia.org/w/index.php?title=File:Flag_of_Yaroslavl_Oblast.svg *License*: unknown *Contributors*: Mikhail Medvedev, Dmitry Ivanov

File:Coat of Arms of Yaroslavl Oblast (2011) full.svg *Source*: http://en.wikipedia.org/w/index.php?title=File:Coat_of_Arms_of_Yaroslavl_Oblast_(2011)_full.svg *License*: unknown *Contributors*: Mikhail Medvedev, Dmitry Ivanov

File:Flag of Moscow (Russia).png *Source*: http://en.wikipedia.org/w/index.php?title=File:Flag_of_Moscow_(Russia).png *License*: unknown *Contributors*: Anime Addict AA, Chesnok, Conscious, Cycn, Homo lupus, INS Pirat, Imz, Little Savage, Oren neu dag, Panther

File:Coat of Arms of Moscow.png *Source*: http://en.wikipedia.org/w/index.php?title=File:Coat_of_Arms_of_Moscow.png *License*: unknown *Contributors*: .:Ajvol:., Bryan Derksen, ChristianBier, Erri4a, EugeneZelenko, Imz, JR98664, Kneiphof, Oren neu dag, Panther, Skipjack, Sputnik, Twthmoses, Wst, Xiong Chiamiov, Yaleks

File:Flag of Saint Petersburg Russia.svg *Source*: http://en.wikipedia.org/w/index.php?title=File:Flag_of_Saint_Petersburg_Russia.svg *License*: unknown *Contributors*: User:Connormah

File:Coat of Arms of Saint Petersburg (2003).png *Source*: http://en.wikipedia.org/w/index.php?title=File:Coat_of_Arms_of_Saint_Petersburg_(2003).png *License*: unknown *Contributors*: User:Avala, User:Deadstar, User:Panther

File:Flag of the Jewish Autonomous Oblast.svg *Source*: http://en.wikipedia.org/w/index.php?title=File:Flag_of_the_Jewish_Autonomous_Oblast.svg *License*: unknown *Contributors*: User:Verdy_p

File:Coat of Arms of Jewish AO.png *Source*: http://en.wikipedia.org/w/index.php?title=File:Coat_of_Arms_of_Jewish_AO.png *License*: unknown *Contributors*: Ahonc, Hellerick, Ilyaroz, Magul, Mattes, Maximaximax, Panther, Wst

File:Flag of Nenets Autonomous District.svg *Source*: http://en.wikipedia.org/w/index.php?title=File:Flag_of_Nenets_Autonomous_District.svg *License*: unknown *Contributors*: User:Conscious

File:Coat of Arms of Nenetsia.png *Source*: http://en.wikipedia.org/w/index.php?title=File:Coat_of_Arms_of_Nenetsia.png *License*: unknown *Contributors*: vector-images.com

File:Flag of Yugra.svg *Source*: http://en.wikipedia.org/w/index.php?title=File:Flag_of_Yugra.svg *License*: unknown *Contributors*: 32X, Aph, Cecil, ChristianBier, Denelson83, Fred the Oyster, Kobac, Mattes, Orange-kun, Permjak, RG72, Scriber en, Smooth O, 3 anonymous edits

File:Coat of Arms of Yugra.svg *Source*: http://en.wikipedia.org/w/index.php?title=File:Coat_of_Arms_of_Yugra.svg *License*: unknown *Contributors*: 32X, Begoon, Kobac, Orange-kun

File:Flag of Chukotka.svg *Source*: http://en.wikipedia.org/w/index.php?title=File:Flag_of_Chukotka.svg *License*: unknown *Contributors*: User:Fibonacci

File:Coat of Arms of Chukotka.svg *Source*: http://en.wikipedia.org/w/index.php?title=File:Coat_of_Arms_of_Chukotka.svg *License*: unknown *Contributors*: User:Agantacroxi

File:Flag of Yamal-Nenets Autonomous District.svg *Source*: http://en.wikipedia.org/w/index.php?title=File:Flag_of_Yamal-Nenets_Autonomous_District.svg *License*: unknown *Contributors*: User:Conscious

File:Coat of Arms of Yamal Nenetsia.svg *Source*: http://en.wikipedia.org/w/index.php?title=File:Coat_of_Arms_of_Yamal_Nenetsia.svg *License*: unknown *Contributors*: User:Cian Akril

File:Russian Subjects merged.png *Source*: http://en.wikipedia.org/w/index.php?title=File:Russian_Subjects_merged.png *License*: unknown *Contributors*: User:Fremantleboy

File:Map of Russia - Arkhangelsk Oblast (2008-03).svg *Source*: http://en.wikipedia.org/w/index.php?title=File:Map_of_Russia_-_Arkhangelsk_Oblast_(2008-03).svg *License*: unknown *Contributors*: User:marmelad

File:Map of Russia - Nenets Autonomous Okrug (2008-03).svg *Source*: http://en.wikipedia.org/w/index.php?title=File:Map_of_Russia_-_Nenets_Autonomous_Okrug_(2008-03).svg *License*: unknown *Contributors*: User:marmelad

File:Map of Russia - Khabarovsk Krai (2008-03).svg *Source*: http://en.wikipedia.org/w/index.php?title=File:Map_of_Russia_-_Khabarovsk_Krai_(2008-03).svg *License*: unknown *Contributors*: User:marmelad

File:Map of Russia - Jewish Autonomous Oblast (2008-03).svg *Source*: http://en.wikipedia.org/w/index.php?title=File:Map_of_Russia_-_Jewish_Autonomous_Oblast_(2008-03).svg *License*: unknown *Contributors*: User:marmelad

File:Map of Russia - Amur Oblast (2008-03).svg *Source*: http://en.wikipedia.org/w/index.php?title=File:Map_of_Russia_-_Amur_Oblast_(2008-03).svg *License*: unknown *Contributors*: User:marmelad

File:Map of Russia - Magadan Oblast (2008-03).svg *Source*: http://en.wikipedia.org/w/index.php?title=File:Map_of_Russia_-_Magadan_Oblast_(2008-03).svg *License*: unknown *Contributors*: User:marmelad

File:Map of Russia - Chukotka Autonomous Okrug (2008-03).svg *Source*: http://en.wikipedia.org/w/index.php?title=File:Map_of_Russia_-_Chukotka_Autonomous_Okrug_(2008-03).svg *License*: unknown *Contributors*: User:marmelad

File:Map of Russia - Irkutsk Oblast (2008-03).svg *Source*: http://en.wikipedia.org/w/index.php?title=File:Map_of_Russia_-_Irkutsk_Oblast_(2008-03).svg *License*: unknown *Contributors*: User:marmelad

File:Map of Russia - Buryat Republic (2008-03).svg *Source*: http://en.wikipedia.org/w/index.php?title=File:Map_of_Russia_-_Buryat_Republic_(2008-03).svg *License*: unknown *Contributors*: User:marmelad

File:Map of Russia - Zabaykalsky Krai (2008-03).svg *Source*: http://en.wikipedia.org/w/index.php?title=File:Map_of_Russia_-_Zabaykalsky_Krai_(2008-03).svg *License*: unknown *Contributors*: User:marmelad

File:Map of Russia - Tyumen Oblast (2008-03).svg *Source*: http://en.wikipedia.org/w/index.php?title=File:Map_of_Russia_-_Tyumen_Oblast_(2008-03).svg *License*: unknown *Contributors*: User:marmelad

File:Map of Russia - Yamalo-Nenets Autonomous Okrug (2008-03).svg *Source*: http://en.wikipedia.org/w/index.php?title=File:Map_of_Russia_-_Yamalo-Nenets_Autonomous_Okrug_(2008-03).svg *License*: unknown *Contributors*: User:marmelad

File:Map of Russia - Khanty-Mansi Autonomous Okrug (2008-03).svg *Source*: http://en.wikipedia.org/w/index.php?title=File:Map_of_Russia_-_Khanty-Mansi_Autonomous_Okrug_(2008-03).svg *License*: unknown *Contributors*: User:marmelad

File:Map of Russia - Novosibirsk Oblast (2008-03).svg *Source*: http://en.wikipedia.org/w/index.php?title=File:Map_of_Russia_-_Novosibirsk_Oblast_(2008-03).svg *License*: unknown *Contributors*: User:marmelad

File:Map of Russia - Omsk Oblast (2008-03).svg *Source*: http://en.wikipedia.org/w/index.php?title=File:Map_of_Russia_-_Omsk_Oblast_(2008-03).svg *License*: unknown *Contributors*: User:marmelad

File:Map of Russia - Tomsk Oblast (2008-03).svg *Source*: http://en.wikipedia.org/w/index.php?title=File:Map_of_Russia_-_Tomsk_Oblast_(2008-03).svg *License*: unknown *Contributors*: User:marmelad

File:Map of Russia - Kemerovo Oblast (2008-03).svg *Source*: http://en.wikipedia.org/w/index.php?title=File:Map_of_Russia_-_Kemerovo_Oblast_(2008-03).svg *License*: unknown *Contributors*: User:marmelad

File:Map of Russia - Altai Republic (2008-03).svg *Source*: http://en.wikipedia.org/w/index.php?title=File:Map_of_Russia_-_Altai_Republic_(2008-03).svg *License*: unknown *Contributors*: User:marmelad

File:Map of Russia - Altai Krai (2008-03).svg *Source*: http://en.wikipedia.org/w/index.php?title=File:Map_of_Russia_-_Altai_Krai_(2008-03).svg *License*: unknown *Contributors*: User:marmelad

File:Map of Russia - St. Petersburg (federal city) (2008-03).svg *Source*: http://en.wikipedia.org/w/index.php?title=File:Map_of_Russia_-_St._Petersburg_(federal_city)_(2008-03).svg *License*: unknown *Contributors*: User:marmelad

File:Map of Russia - Leningrad Oblast (2008-03).svg *Source*: http://en.wikipedia.org/w/index.php?title=File:Map_of_Russia_-_Leningrad_Oblast_(2008-03).svg *License*: unknown *Contributors*: User:marmelad

File:Map of Russia - Moscow (federal city) (2008-03).svg *Source*: http://en.wikipedia.org/w/index.php?title=File:Map_of_Russia_-_Moscow_(federal_city)_(2008-03).svg *License*: unknown *Contributors*: User:marmelad

File:Map of Russia - Moscow Oblast (2008-03).svg *Source*: http://en.wikipedia.org/w/index.php?title=File:Map_of_Russia_-_Moscow_Oblast_(2008-03).svg *License*: unknown *Contributors*: User:marmelad

File:Map of Russia - Yaroslavl Oblast (2008-03).svg *Source*: http://en.wikipedia.org/w/index.php?title=File:Map_of_Russia_-_Yaroslavl_Oblast_(2008-03).svg *License*: unknown *Contributors*: User:marmelad

File:Map of Russia - Kostroma Oblast (2008-03).svg *Source*: http://en.wikipedia.org/w/index.php?title=File:Map_of_Russia_-_Kostroma_Oblast_(2008-03).svg *License*: unknown *Contributors*: User:marmelad

File:Map of Russia - Novgorod Oblast (2008-03).svg *Source*: http://en.wikipedia.org/w/index.php?title=File:Map_of_Russia_-_Novgorod_Oblast_(2008-03).svg *License*: unknown *Contributors*: User:marmelad

File:Map of Russia - Pskov Oblast (2008-03).svg *Source*: http://en.wikipedia.org/w/index.php?title=File:Map_of_Russia_-_Pskov_Oblast_(2008-03).svg *License*: unknown *Contributors*: User:marmelad

File:Map of Russia - Krasnodar Krai (2008-03).svg *Source*: http://en.wikipedia.org/w/index.php?title=File:Map_of_Russia_-_Krasnodar_Krai_(2008-03).svg *License*: unknown *Contributors*: User:marmelad

File:Map of Russia - Republic of Adygea (2008-03).svg *Source*: http://en.wikipedia.org/w/index.php?title=File:Map_of_Russia_-_Republic_of_Adygea_(2008-03).svg *License*: unknown *Contributors*: User:marmelad

File:Map of Russia - Republic of Ingushetia (2008-03).svg *Source*: http://en.wikipedia.org/w/index.php?title=File:Map_of_Russia_-_Republic_of_Ingushetia_(2008-03).svg *License*: unknown *Contributors*: User:marmelad

File:Map of Russia - Chechen Republic (2008-03).svg *Source*: http://en.wikipedia.org/w/index.php?title=File:Map_of_Russia_-_Chechen_Republic_(2008-03).svg *License*: unknown *Contributors*: User:marmelad

File:Russian census 2010 - stationary center.jpg *Source*: http://en.wikipedia.org/w/index.php?title=File:Russian_census_2010_-_stationary_center.jpg *License*: unknown *Contributors*: User:Schekinov Alexey Victorovich

File:RR5514-0072R.png *Source*: http://en.wikipedia.org/w/index.php?title=File:RR5514-0072R.png *License*: unknown *Contributors*: EvilTeeth, Викимонетчик

GNU Free Documentation License Version 1.2, November 2002 Copyright (C) 2000,2001,2002 Free Software Foundation, Inc. 59 Temple Place, Suite 330, Boston, MA 02111-1307 USA Everyone is permitted to copy and distribute verbatim copies of this license document, but changing it is not allowed.

0. PREAMBLE

The purpose of this License is to make a manual, textbook, or other functional and useful document "free" in the sense of freedom: to assure everyone the effective freedom to copy and redistribute it, with or without modifying it, either commercially or noncommercially. Secondarily, this License preserves for the author and publisher a way to get credit for their work, while not being considered responsible for modifications made by others. This License is a kind of "copyleft", which means that derivative works of the document must themselves be free in the same sense. It complements the GNU General Public License, which is a copyleft license designed for free software. We have designed this License in order to use it for manuals for free software, because free software needs free documentation: a free program should come with manuals providing the same freedoms that the software does. But this License is not limited to software manuals; it can be used for any textual work, regardless of subject matter or whether it is published as a printed book. We recommend this License principally for works whose purpose is instruction or reference.

1. APPLICABILITY AND DEFINITIONS

This License applies to any manual or other work, in any medium, that contains a notice placed by the copyright holder saying it can be distributed under the terms of this License. Such a notice grants a world-wide, royalty-free license, unlimited in duration, to use that work under the conditions stated herein. The "Document", below, refers to any such manual or work. Any member of the public is a licensee, and is addressed as "you". You accept the license if you copy, modify or distribute the work in a way requiring permission under copyright law. A "Modified Version" of the Document means any work containing the Document or a portion of it, either copied verbatim, or with modifications and/or translated into another language. A "Secondary Section" is a named appendix or a front-matter section of the Document that deals exclusively with the relationship of the publishers or authors of the Document to the Document's overall subject (or to related matters) and contains nothing that could fall directly within that overall subject. (Thus, if the Document is in part a textbook of mathematics, a Secondary Section may not explain any mathematics.) The relationship could be a matter of historical connection with the subject or with related matters, or of legal, commercial, philosophical, ethical or political position regarding them. The "Invariant Sections" are certain Secondary Sections whose titles are designated, as being those of Invariant Sections, in the notice that says that the Document is released under this License. If a section does not fit the above definition of Secondary then it is not allowed to be designated as Invariant. The Document may contain zero Invariant Sections. If the Document does not identify any Invariant Sections then there are none. The "Cover Texts" are certain short passages of text that are listed, as Front-Cover Texts or Back-Cover Texts, in the notice that says that the Document is released under this License. A Front-Cover Text may be at most 5 words, and a Back-Cover Text may be at most 25 words. A "Transparent" copy of the Document means a machine-readable copy, represented in a format whose specification is available to the general public, that is suitable for revising the document straightforwardly with generic text editors or (for images composed of pixels) generic paint programs or (for drawings) some widely available drawing editor, and that is suitable for input to text formatters or for automatic translation to a variety of formats suitable for input to text formatters. A copy made in an otherwise Transparent file format whose markup, or absence of markup, has been arranged to thwart or discourage subsequent modification by readers is not Transparent. An image format is not Transparent if used for any substantial amount of text. A copy that is not "Transparent" is called "Opaque". Examples of suitable formats for Transparent copies include plain ASCII without markup, Texinfo input format, LaTeX input format, SGML or XML using a publicly available DTD, and standard-conforming simple HTML, PostScript or PDF designed for human modification. Examples of transparent image formats include PNG, XCF and JPG. Opaque formats include proprietary formats that can be read and edited only by proprietary word processors, SGML or XML for which the DTD and/or processing tools are not generally available, and the machine-generated HTML, PostScript or PDF produced by some word processors for output purposes only. The "Title Page" means, for a printed book, the title page itself, plus such following pages as are needed to hold, legibly, the material this License requires to appear in the title page. For works in formats which do not have any title page as such, "Title Page" means the text near the most prominent appearance of the work's title, preceding the beginning of the body of the text. A section "Entitled XYZ" means a named subunit of the Document whose title either is precisely XYZ or contains XYZ in parentheses following text that translates XYZ in another language. (Here XYZ stands for a specific section name mentioned below, such as "Acknowledgements", "Dedications", "Endorsements", or "History".) To "Preserve the Title" of such a section when you modify the Document means that it remains a section "Entitled XYZ" according to this definition. The Document may include Warranty Disclaimers next to the notice which states that this License applies to the Document. These Warranty Disclaimers are considered to be included by reference in this License, but only as regards disclaiming warranties: any other implication that these Warranty Disclaimers may have is void and has no effect on the meaning of this License.

2. VERBATIM COPYING

You may copy and distribute the Document in any medium, either commercially or noncommercially, provided that this License, the copyright notices, and the license notice saying this License applies to the Document are reproduced in all copies, and that you add no other conditions whatsoever to those of this License. You may not use technical measures to obstruct or control the reading or further copying of the copies you make or distribute. However, you may accept compensation in exchange for copies. If you distribute a large enough number of copies you must also follow the conditions in section 3. You may also lend copies, under the same conditions stated above, and you may publicly display copies.

3. COPYING IN QUANTITY

If you publish printed copies (or copies in media that commonly have printed covers) of the Document, numbering more than 100, and the Document's license notice requires Cover Texts, you must enclose the copies in covers that carry, clearly and legibly, all these Cover Texts: Front-Cover Texts on the front cover, and Back-Cover Texts on the back cover. Both covers must also clearly and legibly identify you as the publisher of these copies. The front cover must present the full title with all words of the title equally prominent and visible. You may add other material on the covers in addition. Copying with changes limited to the covers, as long as they preserve the title of the Document and satisfy these conditions, can be treated as verbatim copying in other respects. If the required texts for either cover are too voluminous to fit legibly, you should put the first ones listed (as many as fit reasonably) on the actual cover, and continue the rest onto adjacent pages. If you publish or distribute Opaque copies of the Document numbering more than 100, you must either include a machine-readable Transparent copy along with each Opaque copy, or state in or with each Opaque copy a computer-network location from which the general network-using public has access to download using public-standard network protocols a complete Transparent copy of the Document, free of added material. If you use the latter option, you must take reasonably prudent steps, when you begin distribution of Opaque copies in quantity, to ensure that this Transparent copy will remain thus accessible at the stated location until at least one year after the last time you distribute an Opaque copy (directly or through your agents or retailers) of that edition to the public. It is requested, but not required, that you contact the authors of the Document well before redistributing any large number of copies, to give them a chance to provide you with an updated version of the Document.

4. MODIFICATIONS

You may copy and distribute a Modified Version of the Document under the conditions of sections 2 and 3 above, provided that you release the Modified Version under precisely this License, with the Modified Version filling the role of the Document, thus licensing distribution and modification of the Modified Version to whoever possesses a copy of it. In addition, you must do these things in the Modified Version: A. Use in the Title Page (and on the covers, if any) a title distinct from that of the Document, and from those of previous versions (which should, if there were any, be listed in the History section of the Document). You may use the same title as a previous version if the original publisher of that version gives permission. B. List on the Title Page, as authors, one or more persons or entities responsible for authorship of the modifications in the Modified Version, together with at least five of the principal authors of the Document (all of its principal authors, if it has fewer than five), unless they release you from this requirement. C. State on the Title page the name of the publisher of the Modified Version, as the publisher. D. Preserve all the copyright notices of the Document. E. Add an appropriate copyright notice for your modifications adjacent to the other copyright notices. F. Include, immediately after the copyright notices, a license notice giving the public permission to use the Modified Version under the terms of this License, in the form shown in the Addendum below. G. Preserve in that license notice the full lists of Invariant Sections and required Cover Texts given in the Document's license notice. H. Include an unaltered copy of this License. I. Preserve the section Entitled "History", Preserve its Title, and add to it an item stating at least the title, year, new authors, and publisher of the Modified Version as given on the Title Page. If there is no section Entitled "History" in the Document, create one stating the title, year, authors, and publisher of the Document as given on its Title Page, then add an item describing the Modified Version as stated in the previous sentence. J. Preserve the network location, if any, given in the Document for public access to a Transparent copy of the Document, and likewise the network locations given in the Document for previous versions it was based on. These may be placed in the "History" section. You may omit a network location for a work that was published at least four years before the Document itself, or if the original publisher of the version it refers to gives permission. K. For any section Entitled "Acknowledgements" or "Dedications", Preserve the Title of the section, and preserve in the section all the substance and tone of each of the contributor acknowledgements and/or dedications given therein. L. Preserve all the Invariant Sections of the Document, unaltered in their text and in their titles. Section numbers or the equivalent are not considered part of the section titles. M. Delete any section Entitled "Endorsements". Such a section may not be included in the Modified Version. N. Do not retitle any existing section to be Entitled "Endorsements" or to conflict in title with any Invariant Section. O. Preserve any Warranty Disclaimers. If the Modified Version includes new front-matter sections or appendices that qualify as Secondary Sections and contain no material copied from the Document, you may at your option designate some or all of these sections as invariant. To do this, add their titles to the list of Invariant Sections in the Modified Version's license notice. These titles must be distinct from any other section titles. You may add a section Entitled "Endorsements", provided it contains nothing but endorsements of your Modified Version by various parties--for example, statements of peer review or that the text has been approved by an organization as the authoritative definition of a standard. You may add a passage of up to five words as a Front-Cover Text, and a passage of up to 25 words as a Back-Cover Text, to the end of the list of Cover Texts in the Modified Version. Only one passage of Front-Cover Text and one of Back-Cover Text may be added by (or through arrangements made by) any one entity. If the Document already includes a cover text for the same cover, previously added by you or by arrangement made by the same entity you are acting on behalf of, you may not add another; but you may replace the old one, on explicit permission from the previous publisher that added the old one. The author(s) and publisher(s) of the Document do not by this License give permission to use their names for publicity for or to assert or imply endorsement of any Modified Version.

5. COMBINING DOCUMENTS

You may combine the Document with other documents released under this License, under the terms defined in section 4 above for modified versions, provided that you include in the combination all of the Invariant Sections of all of the original documents, unmodified, and list them all as Invariant Sections of your combined work in its license notice, and that you preserve all their Warranty Disclaimers. The combined work need only contain one copy of this License, and multiple identical Invariant Sections may be replaced with a single copy. If there are multiple Invariant Sections with the same name but different contents, make the title of each such section unique by adding at the end of it, in parentheses, the name of the original author or publisher of that section if known, or else a unique number. Make the same adjustment to the section titles in the list of Invariant Sections in the license notice of the combined work. In the combination, you must combine any sections Entitled "History" in the various original documents, forming one section Entitled "History"; likewise combine any sections Entitled "Acknowledgements", and any sections Entitled "Dedications". You must delete all sections Entitled "Endorsements".

6. COLLECTIONS OF DOCUMENTS

You may make a collection consisting of the Document and other documents released under this License, and replace the individual copies of this License in the various documents with a single copy that is included in the collection, provided that you follow the rules of this License for verbatim copying of each of the documents in all other respects. You may extract a single document from such a collection, and distribute it individually under this License, provided you insert a copy of this License into the extracted document, and follow this License in all other respects regarding verbatim copying of that document.

7. AGGREGATION WITH INDEPENDENT WORKS

A compilation of the Document or its derivatives with other separate and independent documents or works, in or on a volume of a storage or distribution medium, is called an "aggregate" if the copyright resulting from the compilation is not used to limit the legal rights of the compilation's users beyond what the individual works permit. When the Document is included in an aggregate, this License does not apply to the other works in the aggregate which are not themselves derivative works of the Document. If the Cover Text requirement of section 3 is applicable to these copies of the Document, then if the Document is less than one half of the entire aggregate, the Document's Cover Texts may be placed on covers that bracket the Document within the aggregate, or the electronic equivalent of covers if the Document is in electronic form. Otherwise they must appear on printed covers that bracket the whole aggregate.

8. TRANSLATION

Translation is considered a kind of modification, so you may distribute translations of the Document under the terms of section 4. Replacing Invariant Sections with translations requires special permission from their copyright holders, but you may include translations of some or all Invariant Sections in addition to the original versions of these Invariant Sections. You may include a translation of this License, and all the license notices in the Document, and any Warranty Disclaimers, provided that you also include the original English version of this License and the original versions of those notices and disclaimers. In case of a disagreement between the translation and the original version of this License or a notice or disclaimer, the original version will prevail. If a section in the Document is Entitled "Acknowledgements", "Dedications", or "History", the requirement (section 4) to Preserve its Title (section 1) will typically require changing the actual title.

9. TERMINATION

You may not copy, modify, sublicense, or distribute the Document except as expressly provided for under this License. Any other attempt to copy, modify, sublicense or distribute the Document is void, and will automatically terminate your rights under this License. However, parties who have received copies, or rights, from you under this License will not have their licenses terminated so long as such parties remain in full compliance.

10. FUTURE REVISIONS OF THIS LICENSE

The Free Software Foundation may publish new, revised versions of the GNU Free Documentation License from time to time. Such new versions will be similar in spirit to the present version, but may differ in detail to address new problems or concerns. See http://www.gnu.org/copyleft/. Each version of the License is given a distinguishing version number. If the Document specifies that a particular numbered version of this License "or any later version" applies to it, you have the option of following the terms and conditions either of that specified version or of any later version that has been published (not as a draft) by the Free Software Foundation. If the Document does not specify a version number of this License, you may choose any version ever published (not as a draft) by the Free Software Foundation. ADDENDUM: How to use this License for your documents To use this License in a document you have written, include a copy of the License in the document and put the following copyright and license notices just after the title page: Copyright (c) YEAR YOUR NAME. Permission is granted to copy, distribute and/or modify this document under the terms of the GNU Free Documentation License, Version 1.2 or any later version published by the Free Software Foundation; with no Invariant Sections, no Front-Cover Texts, and no Back-Cover Texts. A copy of the license is included in the section entitled "GNU Free Documentation License". If you have Invariant Sections, Front-Cover Texts and Back-Cover Texts, replace the "with...Texts." line with this: with the Invariant Sections being LIST THEIR TITLES, with the Front-Cover Texts being LIST, and with the Back-Cover Texts being LIST. If you have Invariant Sections without Cover Texts, or some other combination of the three, merge those two alternatives to suit the situation. If your document contains nontrivial examples of program code, we recommend releasing these examples in parallel under your choice of free software license, such as the GNU General Public License, to permit their use in free software.

MIX
Papier aus verantwortungsvollen Quellen
Paper from responsible sources
FSC® C105338

Printed by Books on Demand GmbH, Norderstedt / Germany